Study Guide for

The Human Body in
Health and Illness

Study Guide for

The Human Body in Health and Illness

Sixth Edition

Barbara Herlihy, BSN, MA, PhD (Physiology), RN
Professor of Biology
University of the Incarnate Word
School of Mathematics, Science, and Engineering
San Antonio, Texas

ELSEVIER

ELSEVIER

3251 Riverport Lane
St. Louis, Missouri 63043

To all those compassionate souls who provide "forever homes" for our abandoned and abused fur babies. They truly reflect the heart of Meister Eckhart, a 13th-century mystic, who celebrated every creature as an expression of divine wisdom.

Barbara Herlihy

Preface

Questions, questions, and more questions! The *Study Guide for The Human Body in Health and Illness* is designed to help you learn the basic concepts of anatomy and physiology through relentless questioning. Each chapter in the *Study Guide* corresponds to a chapter in the textbook. Some questions are easy and require simple recall; other exercises are more difficult and are designed to help you synthesize and integrate basic concepts. A strategy that I have found very helpful is to ask the same question in several different ways. This requires you to view the content from several different perspectives and encourages you to think critically and to integrate many seemingly unrelated facts. This *Study Guide* will make you work.

It is recommended that you work through all the exercises in every chapter. Working in groups reduces stress, encourages learning, and makes the learning process more enjoyable. More important, student-to-student interaction encourages active learning.

ORGANIZATION

The *Study Guide* chapters are divided into two parts: **Part I, Mastering the Basics**, contains matching, ordering, labeling, diagram reading, similars and dissimilars, and coloring exercises for each content area in the corresponding textbook chapter, helping you learn basic anatomy and physiology knowledge; **Part II, Putting It All Together**, contains multiple-choice practice quizzes, case studies, and puzzles that integrate the chapter content.

Throughout the *Study Guide*, there is a concerted effort to use the medical terminology that was introduced in the textbook and will certainly be used in clinical situations. For instance, words such as *diagnosis, hypokalemia*, and *hyperglycemia* are used frequently and require mastery. As in the textbook, *pathophysiology* is used when it serves to explain the normal anatomy and physiology.

Throughout the *Study Guide*, page references from the textbook are provided to assist you in answering the questions. A complete Answer Key has been provided for your instructor.

Part I: Mastering the Basics

Matching
You are asked to match the words or terms in one column with descriptive terms in a second column.

Ordering
The ordering exercises ask you to arrange a series of events or structures in the correct order. For instance, one exercise is ordering the structures through which blood flows from the right atrium through the heart to the aorta. Other ordering exercises include the sequence of events at the neuromuscular junction, the flow of urine from the kidney through the urethra, and the flow of blood from the finger to the toe.

Labeling and Coloring
Many of the illustrations that appear in the textbook are reproduced in the *Study Guide*. You will need to label the figure and, in some instances, color a particular part of it. Coloring helps to focus your attention on a particularly important anatomic structure. For example, color the right side of the heart blue, indicating unoxygenated blood, and the left side red, indicating oxygenated blood.

Read the Diagram
These exercises ask you to interpret illustrations from the textbook to promote understanding of a particular function or process.

Similars and Dissimilars
Four words are listed; you are asked to identify the word that is least related to the other three words.

Part II: Putting It All Together

Multiple-Choice Questions
Each chapter contains multiple-choice questions. In addition, a simple case study and a series of related multiple-choice questions appear at the end of each chapter.

Groups and Puzzles
The grouping exercise asks you to integrate information by grouping together related topics and excluding unrelated topics. The puzzles, while entertaining, are integrative and instructive. You are asked to eliminate anatomical terms until you discover the answer. A hint appears in the title.

Acknowledgments

As with the text, the creation and publication of this *Study Guide* involved the combined efforts of many people. I want to thank the staff of Elsevier for their efforts; they are talented, beyond competent, and just plain nice. I especially want to thank Elizabeth Kilgore, Apoorva Velpur, Laurie Gower, Deepthi Unni, and Kellie White for their encouragement, persistence, patience, sense of humor, and close attention to detail. Massive amounts of information came together very peacefully!

Many thanks to Jerry for enduring another edition and proofreading miles of words. It cured him from undertaking a text of his own. The same goes to the pet population, which faithfully camped out around my cluttered writing table. And to my grandchildren—not quite so understanding—who just wanted me to play with them. Love to my friend and immediate boss. Dr Bonnie McCormick at the University of the Incarnate Word, kind, patient, and so competent.

Contents

1 Introduction to the Human Body **1**

2 Basic Chemistry **7**

3 Cells **13**

4 Cell Metabolism **19**

5 Microbiology Basics **25**

6 Tissues and Membranes **29**

7 Integumentary System and Body Temperature **35**

8 Skeletal System **41**

9 Muscular System **59**

10 Nervous System: Nervous Tissue and Brain **67**

11 Nervous System: Spinal Cord and Peripheral Nerves **77**

12 Autonomic Nervous System **83**

13 Sensory System **89**

14 Endocrine System **101**

15 Blood **113**

16 Anatomy of the Heart **125**

17 Function of the Heart **135**

18 Anatomy of the Blood Vessels **143**

19 Functions of the Blood Vessels **157**

20 Lymphatic System **165**

21 Immune System **169**

22 Respiratory System **175**

23 Digestive System **183**

24 Urinary System **197**

25 Water, Electrolyte, and Acid–Base Balance **207**

26 Reproductive Systems **215**

27 Human Development and Heredity **223**

1 Introduction to the Human Body

OBJECTIVES

1. Define the terms *anatomy* and *physiology*.
2. List the levels of organization of the human body.
3. Describe the 12 major organ systems.
4. Define homeostasis.
5. Describe the anatomical position.
6. List common terms used for relative positions of the body.
7. Describe the three major planes of the body.
8. List anatomical terms for regions of the body.
9. Describe the major cavities of the body.

Part I: Mastering the Basics

MATCHING

General Terms

Directions: Match the following terms to the most appropriate definition by writing the correct letter in the space provided. Some terms may be used more than once. See text, pp. 1, 6.

A. physiology   C. pathophysiology
B. homeostasis   D. anatomy

1. _____ Branch of science that studies the structure of the body

2. _____ Branch of science that describes how the body functions

3. _____ Branch of science that describes the consequences of improper function of the body (as in disease)

4. _____ Word from the Greek meaning "to dissect"

5. _____ Word describing, for example, body temperature remaining at 37°C (98.6°F), despite the fact that the person is swimming in water that is 22°C (72°F)

6. _____ Word describing the lowering of the blood glucose level to normal after eating a meal

7. _____ Example: the heart has four chambers

8. _____ Example: heart muscle contracts forcefully and pumps blood into the blood vessels

9. _____ Example: a damaged heart muscle pumps an insufficient quantity of blood

10. _____ The femur, located in the thigh, is the largest bone in the body.

Anatomical Terms

Directions: Match the following terms to the most appropriate definition by writing the correct letter in the space provided. Some terms may be used more than once. See text, pp. 6-7.

A. superior   G. anterior
B. medial   H. anatomical position
C. distal   I. lateral
D. inferior   J. posterior
E. proximal   K. deep
F. superficial   L. peripheral

1. _____ The body is standing erect, with the face forward, the arms at the side, and the toes and the palms of the hands directed forward.

2. _____ Part that is above another part or is closer to the head; opposite of *inferior*

3. _____ Toward the front (the belly surface); another word is *ventral*

4. _____ Toward the back surface; another word is *dorsal*

5. _____ Part that is located below another part or is closer to the feet; opposite of *superior*

6. _____ Toward the midline of the body; opposite of *lateral*

7. _____ Structure that is nearer the trunk or main part of the body; opposite of *distal*

8. _____ Part that is located on or near the surface of the body; opposite of *deep*

9. _____ Part that is located away from the center; opposite of *central*

10. _____ Away from the midline of the body; opposite of *medial*

11. _____ Position of the blood vessels relative to the heart (central location)

12. _____ Opposite of *superficial*

13. _____ Opposite of *proximal*

14. _____ The fingers are _____ to the wrist.

15. _____ The elbow is _____ to the wrist.

16. _____ The abdomen is _____ to the chest.

17. _____ The head is _____ to the shoulders.

18. _____ The mouth is _____ to the nose.

READ THE DIAGRAM

Directions: Referring to the diagram, fill in the spaces with the correct letters. Not all letters are used—(L) is left; (R) is right. See text, pp. 6-8.

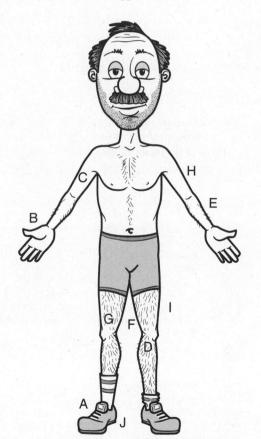

1. _____ Point that is distal to the (L) elbow and proximal to the wrist

2. _____ Point that is proximal to the (R) elbow

3. _____ Point that is immediately distal to the (L) patella

4. _____ Point that is immediately proximal to the (R) patella

5. _____ Point that indicates the lateral aspect of the (R) foot

6. _____ Point that is distal to the (L) hip and proximal to the knee

7. _____ Point that indicates the medial aspect of the right foot

8. _____ Point that is distal to the (L) axillary area and proximal to the antecubital area

XS, YS, ZS, AND CIRCLES

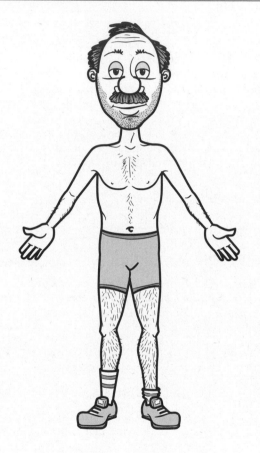

1. Encircle the following areas: cervical, oral, umbilical.

2. Place a string of Xs along the sternal area.

3. Place Ys in the antecubital spaces.

4. Place a string of Zs along the brachial areas.

5. Draw an arrow to the left inguinal area.

6. Place a W on the right patellar area.

7. Place a D on both deltoid areas.

8. Place an A on the medial right thigh.

9. Place a B on a point distal to the left patella and proximal to the left ankle.

10. Place a C on a point proximal to the left elbow.

11. Draw a straight line to create a sagittal plane.

12. Draw a wavy line, creating a transverse plane.

MATCHING

Regional Body Terms

Directions: Match the following terms to the most appropriate definition by writing the correct letter in the space provided. See text, pp. 8-9.

A. digital	L. oral
B. axillary	M. occipital
C. buccal	N. pubic
D. cervical	O. patellar
E. lumbar	P. pedal
F. deltoid	Q. popliteal
G. scapular	R. femoral
H. umbilical	S. brachial
I. antecubital	T. sternal
J. gluteal	U. flank
K. inguinal	

1. _____ Neck region

2. _____ Groin region

3. _____ Navel, or "belly button" area

4. _____ Armpit

5. _____ Kneecap area

6. _____ Between the cheek and gum

7. _____ Pertaining to the mouth

8. _____ Pertaining to the back (posterior) of the head

9. _____ Lower back area, extending from the chest to the hips

10. _____ Where you sit; the buttocks area

11. _____ Front aspect of the elbow area

12. _____ Area behind the knee

13. _____ Shoulder area

14. _____ Breastbone area

15. _____ Genital area

16. _____ Referring to the arm

17. _____ Referring to fingers and toes

18. _____ Foot area

19. _____ Shoulder blade area

20. _____ Area on the sides between the lower ribs and hip

21. _____ Thigh region

MATCHING

Cavities of the Body

Directions: Match the following terms to the most appropriate definition by writing the correct letter in the space provided. Some terms may be used more than once. See text, pp. 9-11.

A. dorsal cavity	E. ventral cavity
B. cranial cavity	F. abdominopelvic cavity
C. vertebral cavity	
D. thoracic cavity	G. pleural cavity

1. _____ Cavity that is located in the skull and contains the brain

2. _____ Cavity that extends from the cranial cavity; contains the spinal cord

3. _____ Also called the *spinal canal*

4. _____ Cavity that is located in the front of the body; contains the thoracic cavity and the abdominopelvic cavity

5. _____ Cavity that is divided into quadrants

6. _____ Cavity that is located toward the back of the body; contains the cranial cavity and the vertebral cavity

7. _____ Cavity that is divided into nine regions

8. _____ Lower ventral cavity that is separated from the thoracic cavity by the diaphragm

9. _____ Part of the ventral cavity that contains the mediastinum

10. _____ Part of the thoracic cavity that contains the lungs

11. _____ Ventral cavity that is inferior to the thoracic cavity

12. _____ Ventral cavity that contains the liver, stomach, spleen, and intestines

13. _____ Cavity that contains the brain and spinal cord

14. _____ Cavity that contains the heart and lungs

15. _____ Cavity that is described in quadrants (RUQ, LUQ, RLQ, LLQ)

READ THE DIAGRAM

Directions: Indicate the letter on the diagram that is described below—(L) is left; (R) is right. Not all letters are used. Some letters may be used more than once.

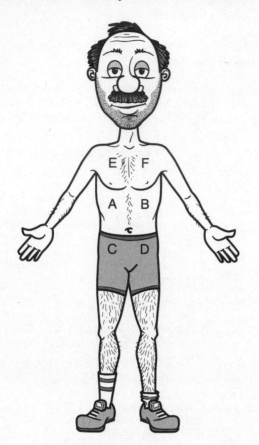

1. _____ The point that is inferior to the diaphragm and located in the left lower quadrant (LLQ)

2. _____ The point over the left pleural cavity

3. _____ The point that is in both the right upper quadrant (RUQ) and hypochondriac region (R)

4. _____ The point on the right side of the body that is superior to the umbilicus and inferior to the diaphragm

5. _____ The point of pain for acute appendicitis—right lower quadrant (RLQ)

6. _____ The point on the left side of the body that does not "fall into" a quadrant

7. _____ The point that is inferior to the RUQ

8. _____ The point that is superior to the RLQ

SIMILARS AND DISSIMILARS

Directions: Circle the word in each group that is least similar to the others. Indicate the similarity of the three words on the line below each question.

1. stomach   ovary   sagittal   heart
_____

2. heart   thoracic   ventral   pleural
_____

3. superior   anterior   pleural   deep
_____

4. lungs   dorsal   thoracic   pelvic
_____

5. RUQ   cranial   umbilical   left inguinal
_____

6. LLQ   left hypochondriac   RLQ   thoracic
_____

7. umbilical   epigastric   right inguinal   cranial
_____

8. urinary   reproductive   stomach   nervous
_____

9. ventral   diaphragm   dorsal   abdominopelvic
_____

10. proximal   flank   superior   distal
_____

11. thoracic   sagittal   frontal   transverse
_____

12. ventral   cranial   spinal   vertebral
_____

Part II: Putting It All Together

MULTIPLE CHOICE

Directions: Choose the correct answer.

1. The heart and blood vessels work together to pump blood throughout the body. What is the word that describes the heart and blood vessels?
   a. tissue
   b. molecule
   c. organ
   d. organ system

2. The inguinal region
   a. is in the thoracic cavity.
   b. houses the heart.
   c. refers to the groin.
   d. is superior to the umbilical region.

3. The sternum (breastbone) is
   a. inferior to the umbilicus.
   b. deep to the lungs.
   c. superficial to the heart.
   d. posterior to the heart.

4. Which word describes the route of administration of a drug given by mouth?
   a. cranial
   b. inguinal
   c. oral
   d. antecubital

5. The appendix is located in the RLQ of which cavity?
   a. dorsal
   b. cranial
   c. abdominopelvic
   d. spinal

6. The liver is located in the RUQ and the appendix is located in the RLQ. Which of the following describes the position of the liver relative to the appendix?
   a. distal
   b. superior
   c. anterior
   d. deep

7. Which region of the abdomen surrounds the navel (belly button)?
   a. epigastric region
   b. popliteal area
   c. umbilical region
   d. right iliac region

8. Which of the following regions is included within the RUQ?
   a. umbilical
   b. hypogastric
   c. right iliac
   d. right hypochondriac

9. Which of the following describes the division of the body by a transverse plane?
   a. a front and a back
   b. a left and a right
   c. a top and a bottom
   d. medial and lateral

10. Which of the following describes the division of the body by a coronal plane?
    a. a front and a back
    b. a top and a bottom
    c. a left and a right
    d. internal and external

11. Which of the following describes the division of the body by a sagittal plane?
    a. a front and a back
    b. a left and a right
    c. a top and a bottom
    d. dorsal and ventral

12. Which of the following divides the body into a front and a back?
    a. the frontal plane
    b. the sagittal plane
    c. a cross section
    d. transverse plane

13. Which of the following describes the position of the wrist relative to the elbow?
    a. proximal
    b. superior
    c. dorsal
    d. distal

14. Which of the following describes the position of the antecubital space relative to the carpal region?
    a. proximal
    b. medial
    c. distal
    d. deep

15. Which of the following describes the position of the wrist relative to the fingers?
    a. proximal
    b. inferior
    c. ventral
    d. distal

16. The mediastinum is contained within which cavity?
    a. dorsal
    b. abdominal
    c. pelvic
    d. thoracic

5

17. Which of the following describes the location of the patella (kneecap) relative to the ankle?
    a. inferior
    b. distal
    c. lateral
    d. proximal

18. Which group is incorrect?
    a. planes: transverse, sagittal, frontal
    b. ventral cavities: thoracic, abdominopelvic, spinal
    c. cavities: dorsal, ventral
    d. organs: heart, stomach, lungs

19. Which group is incorrect?
    a. cavities: dorsal, ventral
    b. ventral cavities: thoracic, abdominopelvic, pleural
    c. organ systems: circulatory, digestive, respiratory, immune
    d. organs: heart, stomach, lungs, mediastinum

20. Which group is incorrect?
    a. planes: transverse, sagittal, frontal
    b. cavities: dorsal, ventral
    c. dorsal cavities: thoracic, abdominopelvic
    d. organ systems: skeletal, digestive, respiratory, immune

PUZZLE

Hint: Navel-Gazing Territory

Directions: Perform the following functions on the Sequence of Words that follows. When all the functions have been performed, you are left with a word or words related to the hint. Record your answer in the space provided.

Functions: Remove the following:

1. Four quadrants

2. Word for the groin area

3. Words (three) that refer to parts of the upper extremities

4. Words (four) that refer to parts of the lower extremities

5. Cavities (three) of the ventral cavity

6. Cavities (two) of the dorsal cavity

7. Word for the lower back region

8. Membranes (two) that have a visceral and parietal layer

9. Planes of the body (three)

10. Word paired with distal

11. Word paired with posterior

12. Word paired with superior

Sequence of Words

CORONALPROXIMALBRACHIALCRANIALIN
GUINALLUMBARABDOMINOPELVICTHO
RACICPERITONEUMTRANSVERSEPATELLAR
RUQDIGITALFEMORALPLEURALUQANTERI
ORRLQUMBILICALREGIONPEDALANTECU
BITALSAGITTALINFERIORSPINALLLQPLEU
RALPOPLITEAL

Answer: _____

2 Basic Chemistry

OBJECTIVES

1. Define the terms *matter, element,* and *atom,* and do the following:
   - List the four elements that comprise 96% of body weight.
   - Describe the three components of an atom.
   - Describe the role of electrons in the formation of chemical bonds.
2. Differentiate among ionic, covalent, and hydrogen bonds.
3. Explain ions, including the differences among electrolytes, cations, and anions.
4. Explain the difference between a molecule and a compound, and list five reasons why water is essential to life.
5. Explain the role of catalysts and enzymes.
6. Differentiate between an acid and a base, and define pH.
7. List the six forms of energy, and describe the role of adenosine triphosphate (ATP) in energy transfer.
8. Differentiate among a mixture, solution, suspension, colloidal suspension, and precipitate.

Part I: Mastering the Basics

MATCHING

Matter, Elements, and Atoms

Directions: Match the following words and symbols to the most appropriate definition by writing the correct letter in the space provided. Some words and symbols may be used more than once. See text, pp. 15-17.

A. chemistry	G. Na
B. matter	H. Ca
C. element	I. O
D. atom	J. N
E. K	K. H
F. Cl	L. Fe

1. _____ A fundamental substance that cannot be broken down into a simpler form by ordinary chemical means

2. _____ Smallest unit of an element that has that element's characteristics

3. _____ Anything that occupies space and has weight

4. _____ Symbol for iron

5. _____ Composed of three particles: protons, neutrons, and electrons

6. _____ Exists in three states: solid, liquid, and gas

7. _____ The study of matter

8. _____ Symbol for oxygen

9. _____ Symbol for sodium

10. _____ Symbol for nitrogen

11. _____ Symbol for potassium

12. _____ Symbol for hydrogen

13. _____ Symbol for calcium

14. _____ Symbol for chlorine

15. _____, _____ Identify the two atoms in table salt.

16. _____, _____ Identify the two atoms in potassium chloride.

MATCHING

The Atom

Directions: Match the following terms to the most appropriate definition by writing the correct letter in the space provided. Some terms may be used more than once. See text, pp. 16-17.

A. electron(s)	E. atomic mass
B. proton(s)	F. radioisotope
C. neutron	G. radioactivity
D. atomic number	H. isotope

1. _____ Number of protons in the nucleus

2. _____ Sum of the protons and the neutrons

3. _____ Helium has two protons and two neutrons; this is what the number 2 indicates.

4. _____ Helium has two protons and two neutrons; this is what the number 4 indicates.

5. _____ Carries a negative charge and circulates in orbits around the nucleus

6. _____ Carries a positive charge and is located within the nucleus

7. _____ Has a neutral charge and is located within the nucleus /

8. _____ In each atom, the number of these is equal to the number of protons.

9. _____ Different form of the same element (same atomic number but a different atomic mass); an example is "heavy hydrogen"

10. _____ Unstable isotope

11. _____ Spontaneous decay of a radioisotope

12. _____ The atomic number is determined by the number of _____.

13. _____ These atomic particles are represented by the planets encircling the sun in Fig. 2.2.

14. _____ This particle is added or removed in making an isotope of an atom.

15. _____, _____ These particles are represented by the sun in Fig. 2.2.

MATCHING

Bonds

Directions: Match the following words to the most appropriate definition by writing the correct letter in the space provided. Some words may be used more than once. See text, pp. 18-19.

A. ionic bond
B. covalent bond
C. hydrogen bond

1. _____ Type of bond formed when electrons are shared by atoms

2. _____ Type of bond that forms between water molecules

3. _____ Type of bond that forms water, H_2O

4. _____ Type of bond between sodium and chloride in table salt, NaCl

5. _____ Intermolecular bond

6. _____ Type of bond formed when one atom donates an electron to another atom

7. _____ Type of bond usually formed when carbon interacts with another atom

MATCHING

Cations, Anions, and Electrolytes

Directions: Match the following words to the most appropriate definition by writing the correct letter in the space provided. Some words may be used more than once. See text, p. 20.

A. cation(s)
B. anion(s)
C. electrolyte(s)
D. ion(s)
E. ionization

1. _____ Atom that carries an electrical charge

2. _____ Sodium ion

3. _____ Chloride ion

4. _____ Formed as electrons are lost or gained

5. _____ Classification of NaCl

6. _____ Positively charged ion

7. _____ Negatively charged ion

8. _____, _____ Ions represented as Na^+, K^+, and Ca^{2+}

9. _____ Dissociation of NaCl into Na^+ and Cl^-

10. _____ Substance that can ionize

11. _____ Cl^-, Na^+, K^+, and Ca^{2+}

MATCHING

Molecules and Compounds

Directions: Match the following words to the most appropriate definition by writing the correct letter in the space provided. Some words may be used more than once. See text, pp. 20-22.

A. water
B. oxygen
C. carbon dioxide
D. chemical reaction
E. catalyst
F. molecule(s)
G. compound(s)

1. _____ Classification of O_2 and N_2

2. _____ Substances that contain molecules formed by two or more different atoms

3. _____ Classification of H_2O in addition to molecule

4. _____ Most abundant compound in the body

5. _____ Molecule that exists in nature as a gas and plays an essential metabolic role in supplying the cells of the body with energy

6. _____ Compound is a waste product that is formed when food is chemically broken down for energy

7. _____ This molecule is the reason why cardiopulmonary resuscitation (CPR) must be started immediately.

8. _____ Compound that is considered to be the universal solvent

9. _____ Compound that has the ability to absorb large amounts of heat without itself increasing dramatically in temperature

10. _____ Describes, for example, glucose + O_2 → CO_2 + H_2O + energy

11. _____ Describes the role of an enzyme that increases the rate of a chemical reaction

MATCHING

Acids and Bases

Directions: Match the following words and symbols to the most appropriate definition by writing the correct letter in the space provided. Some words and symbols may be used more than once. See text, pp. 22-24.

A. acid or acidic   C. buffer
B. base or basic   D. pH

1. _____ A scale ranging from 0 to 14 that measures how many H^+ (hydrogen ions) are in solution

2. _____ Electrolyte that dissociates into H^+ and an anion

3. _____ Substance that removes H^+ from solution

4. _____ Describes a pH of 7.6

5. _____ Describes the effect of an antacid on stomach H^+

6. _____ Turns litmus paper blue

7. _____ Also referred to as *alkaline*

8. _____ Chemical substance that prevents large changes in pH

9. _____ Describes normal pH of urine

10. _____ Describes normal pH of blood

11. _____ Describes normal pH of gastric (stomach) juice

12. _____ Substance that can either donate or remove H^+ from solution

13. _____ Turns litmus paper pink

READ THE DIAGRAM

pH Scale

Directions: Referring to Fig. 2.6 in the textbook, write the numbers from the pH scale in the spaces below. See text, pp. 23-24.

1. _____ Which number indicates a neutral pH?

2. _____ What is the acidic range?

3. _____ What is the basic range?

4. _____ What is the alkaline range?

5. _____ Relative to pH 7, which numbers indicate a higher concentration of H^+?

6. _____ Relative to pH 7, which numbers indicate a lower concentration of H^+?

7. _____ Range for blood pH

8. _____ Range for intestinal contents

9. _____ Range for stomach contents

10. _____ Range for urine

MATCHING

Energy

Directions: Match the following words and symbols to the most appropriate definition by writing the correct letter in the space provided. Some words and symbols may be used more than once. See text, pp. 24-25.

A. mechanical   E. electrical
B. thermal   F. radiant
C. nuclear   G. ATP
D. chemical

1. _____ A log is burned, providing light, as chemical energy is converted into this type of energy.

2. _____ Walking is an expression of this type of energy.

3. _____ A log is burned, warming everyone around the campfire, as chemical energy is converted into this type of energy.

4. _____ The heart pushes blood into large blood vessels as chemical energy is converted into this type of energy.

5. _____ Type of energy that holds atoms together

6. _____ Type of energy that is released from the movement of ions across cell membranes

7. _____ Energy transfer substance

8. _____ The unstable nucleus of an isotope spontaneously decays, thereby emitting this type of energy.

9. _____ Responsible for body temperature as chemical energy is converted to this type of energy

10. _____ Example: sugar $\rightarrow CO_2 + H_2O$ + energy (ATP)

MATCHING

Mixtures, Solutions, Suspensions, and Precipitates

Directions: Match the following words to the most appropriate definition by writing the correct letter in the space provided. Some words may be used more than once. See text, pp. 25-27.

A. mixture
B. solution
C. suspension(s)
D. aqueous
E. colloidal suspension
F. tincture
G. precipitate

1. _____ An example is blood plasma because the proteins remain suspended within the plasma.

2. _____ Solution in which water is the solvent

3. _____ Solution in which alcohol is the solvent

4. _____ Combinations of two or more substances that can be separated by ordinary physical means

5. _____ Examples include mayonnaise, egg white, and jellies.

6. _____ Mixture that contains a solvent and solute; there is an even distribution of the solute with the solution

7. _____ Combination of sugar and little bits of iron

8. _____ Example: the undesirable and dangerous white flakes that occasionally form when you add a drug to an IV salt solution

9. _____ Mixture that must be shaken to prevent settling of particles

10. _____ Suspension in which the particles are so small that they do not need to be shaken to keep them evenly distributed

11. _____ The solid formed in a solution during a chemical reaction

12. _____ Example is sea or salt water

SIMILARS AND DISSIMILARS

Directions: Circle the word in each group that is least similar to the others. Indicate the similarity of the three words on the line below each question.

1. molecules   protons   electrons   neutrons

   _____

2. Na ion   K ion   Cl ion   H ion

   _____

3. ionic   covalent   pH   intermolecular

   _____

4. anion   neutron   ion   cation

   _____

5. H^+ acidic   pH < 6.8   alkalosis

   _____

6. basic   pH > 7.6   potassium   alkaline

   _____

7. Mg   N H_2O Zn

   _____

8. chemical   pH   mechanical   thermal

   _____

9. aqueous   water   tincture   ideal solvent

   _____

Part II: Putting It All Together

MULTIPLE CHOICE

Directions: Choose the correct answer.

1. Which of the following describes activities such as chewing food and chopping a log?
   a. chemical change
   b. neutralization reaction
   c. ionization reaction
   d. physical change

2. Which of the following is a strong acid?
   a. NaOH
   b. H_2O
   c. blood
   d. HCl

3. Which of the following is/are classified as thermal, chemical, or radiant?
   a. cations
   b. anions
   c. energy
   d. electrolytes

4. Which of the following words best describes a radio-isotope?
   a. acidic
   b. alkaline
   c. unstable
   d. ionization

5. Which of the following describes the chlorine atom when its outer electron shell gains one electron?
   a. +1 positive charge
   b. no electrical charge
   c. cation
   d. anion

6. Hydrogen has one proton, zero neutrons, and one electron. Which statement is true?
   a. The atomic number is 2.
   b. The atomic mass is 2.
   c. The atomic number is 1.
   d. This is an isotope of helium because it has no neutrons.

7. What type of bond is formed when two hydrogen atoms and one oxygen atom unite to form water?
   a. ionic
   b. hydrogen
   c. intermolecular
   d. covalent

8. What type of reaction occurs when HCl is mixed with an NaOH solution to form a salt (NaCl) and water?
   a. neutralization
   b. agglutination
   c. differentiation
   d. catabolism

9. What is the pH range of blood?
   a. 4.75 to 5.50
   b. 8.45 to 8.95
   c. 7.35 to 7.45
   d. 7.00 to 7.35

10. Which of the following is an electrolyte?
    a. potassium chloride
    b. glucose
    c. water
    d. mayonnaise

11. Which of the following is most acidic?
    a. intestinal contents
    b. blood
    c. pH 7.2
    d. pH 6.6

12. The sodium ion is a(n)
    a. electrolyte.
    b. anion.
    c. compound.
    d. cation.

13. A blood pH of 7.2
    a. has fewer H^+ ions than normal blood pH.
    b. is acidosis.
    c. is within normal limits.
    d. is alkalosis.

14. A blood pH of 7.55
    a. has fewer H^+ ions than normal blood pH.
    b. is within normal limits.
    c. is considered acidotic.
    d. can be lowered by the removal of H^+.

15. When placed in water, sodium chloride (table salt)
    a. lowers the pH of the solution.
    b. forms an NaCl precipitate.
    c. ionizes into Na^+ and Cl^-.
    d. forms a strong acid.

16. Which group is incorrect?
    a. parts of an atom: proton, neutron, electron
    b. electrolytes: NaCl, CaCl$_2$, KCl
    c. chemical bonds: ionic, covalent, intermolecular
    d. cations: sodium ion, potassium ion, chloride ion

17. Which group is incorrect?
    a. parts of an atom: proton, neutron, electron
    b. cations: sodium ion, potassium ion, calcium ion
    c. states of matter: solid, liquid, gas
    d. trace elements: copper, calcium, hydrogen

18. Which group is incorrect?
    a. chemical bonds: ionic, covalent, intermolecular
    b. ions: cations, anions
    c. parts of an atom: protons, neutrons, electrons
    d. carriers of a negative charge: electrons, anions, neutrons

BODY TOON

Hint: Spray Painting the Town

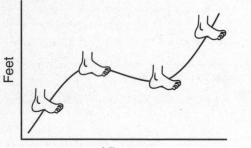

Feet

Minutes

PUZZLE

Hint: YUM . . .

Directions: Perform the following functions on the Sequence of Words that follows. When all the functions have been performed, you are left with a word or words related to the hint. Record your answer in the space provided.

Functions: Remove the following:

1. Clinical condition characterized by pH < 7.35

2. Clinical condition characterized by pH > 7.45

3. Parts (three) of an atom

4. Color of an acid on the pH scale

5. Color of a base on the pH scale

6. Another word for basic

7. Types of bonds (two): sharing and donating

8. Most common compound in the body

9. Energy transfer molecule

10. Name of a water solution and an alcohol solution

11. Names for these ions: Na$^+$, K$^+$, and HCO$_3^-$

12. Different form of the same element (same atomic number but different atomic mass)

13. Chemical substance that can ionize (e.g., NaCl)

Sequence of Words

ISOTOPEWATERELECTROLYTESODIU
MATPALKALINECOVALENTALKALOSISELEC
TRONSPINKPROTONSNEUTRONSBICARBON
ATEAQUEOUSIONICBLUEACIDOSISTINCTUR
ENACHOSPOTASSIUM

Answer: _____

Answer: graph-feetie (graffiti)

3 Cells

OBJECTIVES

1. Label a diagram of the main parts of a typical cell, and do the following:
   ■ Explain the role of the nucleus.
   ■ Describe the functions of the main organelles of the cell.
   ■ Identify the components of the cell membrane.
2. Do the following regarding transport mechanisms:
   ■ Describe the active and passive movements of substances across a cell membrane.
   ■ Define tonicity and compare isotonic, hypotonic, and hypertonic solutions.
3. Describe the phases of the cell cycle, including mitosis.
4. Explain what is meant by cell differentiation.
5. Explain the processes and consequences of uncontrolled and disorganized cell growth and apoptosis.

Part I: Mastering the Basics

MATCHING

Parts of a Typical Cell

Directions: Match the following terms to the most appropriate definition by writing the correct letter in the space provided. Some terms may be used more than once. See text, pp. 30-35.

A. mitochondria
B. nucleus
C. microtubules
D. cilia
E. ribosomes
F. lysosomes
G. flagellum
H. centrioles
I. nuclear membrane
J. cytoplasm
K. Golgi apparatus
L. rough endoplasmic reticulum (RER)
M. smooth endoplasmic reticulum (SER)
N. cell membrane
O. cytosol

1. _____ Control center of the cell; contains most of the DNA

2. _____ Slipper-shaped organelles that produce most of the energy (ATP)

3. _____ Puts the finishing touches on the protein and packages it for export from the cell

4. _____ Structure that separates the nucleus from the cytoplasm

5. _____ Sandpaper-like structure dotted with ribosomes; concerned with protein synthesis

6. _____ Long hairlike projection on the external surface of the cell membrane, such as the tail of the sperm

7. _____ Consists of the cytosol and the organelles

8. _____ Selectively permeable structure that separates intracellular material from extracellular material

9. _____ Short hairlike projections on the outer surface of the cell

10. _____ Digestive organelles that "clean house" within the cell

11. _____ Organelles that help maintain the shape of the cell and assist the cell with movement

12. _____ Gel-like part of the cytoplasm

13. _____ Organelles that either are bound to the endoplasmic reticulum or are free in the cytoplasm; concerned with protein synthesis

14. _____ Type of endoplasmic reticulum concerned with the synthesis of lipids and steroids; does not contain ribosomes

15. _____ Rod-shaped structures that play a key role in cellular reproduction

16. _____ Called the power plants of the cells

The Typical Cell

Directions: Refer to the diagram and fill in the numbers in the spaces below. See text, pp. 30-35.

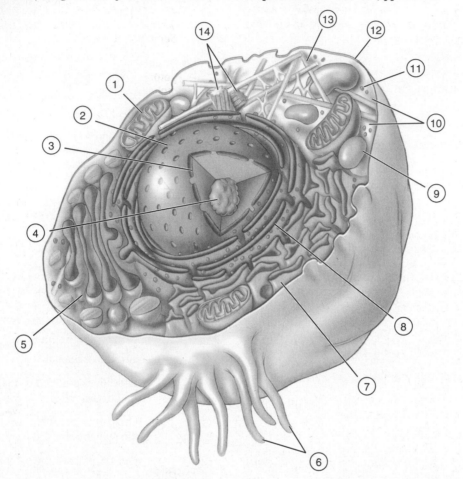

1. _____ Slipper-shaped organelles that produce ATP; called the power plants of the cell

2. _____ Puts the finishing touches on the protein and packages it for export from the cell

3. _____ Sandpaper-like structure dotted with ribosomes; concerned with protein synthesis

4. _____ Selectively permeable structure that separates intracellular material from extracellular material

5. _____ Short hairlike projections on the outer surface of the cell

6. _____ Digestive organelles that "clean house" within the cell

7. _____ Organelles that maintain the shape of the cell and assist the cell with movement

8. _____ Gel-like substance inside the cell but outside the nucleus

9. _____ Type of endoplasmic reticulum concerned with the synthesis of lipids and steroids; does not contain ribosomes

10. _____ Rod-shaped structures that play a key role in cellular reproduction

BODY TOON

Hint: Enzyme Activity

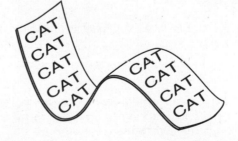

Answer: catalyst (cat-a-list)

Transport Mechanisms

Directions: Match the following terms to the most appropriate definition by writing the correct letter in the space provided. Some terms may be used more than once. See text, pp. 36-41.

A. osmosis
B. diffusion
C. facilitated diffusion
D. phagocytosis
E. exocytosis
F. active transport pump
G. pinocytosis
H. filtration

1. _____ Most commonly used transport mechanism

2. _____ Transport mechanism in which water diffuses from an area where there is more water to an area where there is less water; solute cannot diffuse

3. _____ A pressure gradient is the driving force for this type of transport.

4. _____ Transport mechanism that engulfs a solid particle by the cell membrane; a type of endocytosis

5. _____ Transport mechanism that requires an input of energy to move molecules from an area of lower concentration to an area of higher concentration

6. _____ Movement of a substance from an area of higher concentration to an area of lower concentration

7. _____ Passive transport mechanism in which glucose is helped across the cell membrane by a helper molecule

8. _____ Intake of liquid droplets by the cell membrane; also called *cellular drinking*

9. _____ Example of this transport mechanism is the swelling of a blood clot as water is pulled into the clot.

10. _____ Describes this type of transport mechanism: the blood pressure pushes water and dissolved solute out of the capillaries into the tissue spaces

11. _____ A lysosome eats or ingests a bacterium

12. _____ A protein-containing vesicle within a cell fuses with the cell membrane and ejects the protein.

13. _____ Transport mechanism needed to move additional potassium into the cell if the intracellular concentration of potassium is higher than the extracellular concentration of potassium

Tonicity

Directions: Match the following terms to the most appropriate definition by writing the correct letter in the space provided. Some terms may be used more than once. See text, pp. 39-40.

A. isotonic
B. hypotonic
C. hemolysis
D. hypertonic
E. crenation

1. _____ Shrinking of red blood cells

2. _____ Bursting of red blood cells

3. _____ Solution that is more dilute than the inside of the cell

4. _____ Solution with the same concentration as the solution to which it is compared

5. _____ Solution that is more concentrated than inside of the cell

6. _____ Solution that causes crenation of a red blood cell

7. _____ Solution that causes the red blood cell to swell and burst

8. _____ Normal saline

9. _____ Describes a 10% dextrose solution relative to plasma if a 5% dextrose solution is isotonic to plasma

10. _____ Describes pure water relative to plasma

Cell Division

Directions: Match the following terms to the most appropriate definition by writing the correct letter in the space provided. See text, pp. 41-43.

A. cell cycle
B. cancer
C. mitosis
D. metastasis
E. interphase
F. meiosis
G. G_0
H. stem cell

1. _____ M phase of the cell cycle

2. _____ Ability of cancer cells to spread to distant sites

3. _____ Consists of two phases: interphase and mitosis

4. _____ Cell that can specialize into another type, such as blood cell, nerve cell, muscle cell

5. _____ Type of cell division involved in the body's growth and repair

15

6. _____ Cells stop cycling when they enter this phase.

7. _____ Includes G_1, S, G_2, and M

8. _____ Undifferentiated or unspecialized cell

9. _____ Replication of DNA occurs during this phase of the cell cycle

10. _____ Type of cell division that occurs in sex cells

11. _____ A drug is labeled cell cycle M phase-specific; affects this phase of the cell cycle

12. _____ Phases: prophase, metaphase, anaphase, and telophase

13. _____ Malignant neoplasm

SIMILARS AND DISSIMILARS

Directions: Circle the word in each group that is least similar to the others. Indicate the similarity of the three words on the line below each question.

1. nucleus   mitochondria   melanoma   lysosomes

_____

2. nucleus   ribosomes   DNA   control center

_____

3. cilia   hairlike   flagellum   sandpaper-like

_____

4. ATP   mitochondria   diffusion   energy

_____

5. smooth   cell cycle   endoplasmic   RER

_____

6. diffusion   filtration   active transport   osmosis   pump

_____

7. endocytosis   pinocytosis   filtration   phagocytosis

_____

8. G_1 phase G_2 phase   M phase   S phase

9. prophase   interphase   metaphase   anaphase

10. hypertonic   isotonic   filtration   hypotonic

_____

11. mitochondrion   protein   cristae   power
                    synthesis   plant

_____

12. power plant   RER   protein   amino
                         synthesis   acids

_____

13. pressure gradient   ATP   filtration   passive

_____

14. ribosome   cell membrane   lipid   plasma
                              bilayer   membrane

_____

Part II: Putting It All Together

MULTIPLE CHOICE

Directions: Choose the correct answer.

1. Which of the following is least characteristic of facilitated diffusion?
   a. passive transport
   b. "helper" molecule
   c. solute diffuses down its concentration gradient
   d. pumps solute from an area of lower concentration to an area of higher concentration

2. What is the process that uses energy to move a solute from an area of lower concentration to an area of higher concentration?
   a. diffusion
   b. facilitated diffusion
   c. osmosis
   d. active transport pump

3. Which of the following are located on the cell membrane?
   a. cilia and mitochondria
   b. microvilli and cilia
   c. flagellum and centrioles
   d. Golgi apparatus and RER

4. Perfume the skunk does "his thing." Which of the following words best indicates why you quickly become aware of Perfume's presence?
   a. diffusion
   b. Na/K pump
   c. active transport
   d. osmosis

5. Differentiation is
   a. a type of cell division.
   b. the process that refers to the specialization of cells.
   c. a type of passive transport.
   d. a form of active transport.

6. Which of the following best describes the cell membrane?
   a. nonselective
   b. selectively permeable
   c. impermeable
   d. exclusively lipid-soluble

7. Ribosomes are
   a. only found attached to the endoplasmic reticulum.
   b. located within the nucleus.
   c. concerned with protein synthesis.
   d. the power plants of the cell.

8. Which of the following terms is most related to the mitochondrion?
   a. mucus-secreting
   b. mRNA, tRNA
   c. protein synthesis
   d. energy-producing

9. Which of the following is most related to lysosomes?
   a. protein synthesis
   b. DNA
   c. bound and free
   d. "housecleaning"

10. Prophase, metaphase, anaphase, and telophase
    a. are stages of mitosis.
    b. are the resting phases of the cell cycle.
    c. occur during G_0 of the cell cycle.
    d. are transport mechanisms.

11. Interphase and mitosis
    a. refer to the resting phase of the cell cycle.
    b. are two phases of the cell cycle.
    c. do not include prophase, metaphase, anaphase, or telophase.
    d. are characteristic only of stem cells.

12. G_1, S, and G_2
    a. are stages of mitosis.
    b. occur during interphase.
    c. occur only in stem cells.
    d. occur only in cancerous cells.

13. The ribosome-containing membranous structure
    a. is called the rough endoplasmic reticulum.
    b. is concerned with protein synthesis.
    c. forms intracellular channels that guide the movement of protein.
    d. All of the above are true.

14. Which of the following is a true statement?
    a. The cell membrane is semipermeable, meaning that it allows only for the diffusion of water.
    b. The cytoplasm contains the cytosol and organelles.
    c. Most ATP is made in the nucleus.
    d. The Golgi apparatus is classified as smooth and rough.

15. What is the underlying cause of the cellular effects of aging?
    a. increased numbers of organelles
    b. damage to DNA
    c. increased rate of cellular mitosis
    d. cellular shrinking

16. In a beaker, compartment A, has a 25% glucose solution and compartment B has a 5% glucose solution. The membrane is permeable to both solute and solvent. What transport mechanism causes glucose to move from A to B?
    a. active transport
    b. osmosis
    c. diffusion
    d. filtration

17. Refer to the beaker compartments in the previous question. What transport mechanism causes water to move from B to A?
    a. active transport
    b. osmosis
    c. diffusion
    d. filtration

18. In a beaker, compartment A has a 25% glucose solution and compartment B has a 5% glucose solution. The membrane is permeable to the solvent but not to the solute. What transport mechanism causes water to move from B to A?
    a. filtration
    b. osmosis
    c. active transport
    d. hemolysis

19. In a beaker, compartment A has a 25% glucose solution, and compartment B has a 5% glucose solution. The membrane is permeable to both the solvent and solute. What transport mechanism can move glucose from B to A?
    a. active transport
    b. diffusion
    c. facilitated diffusion
    d. osmosis

20. The pressures within a capillary are as follows: capillary hydrostatic pressure of 35 mm Hg at the arterial end of the capillary and a pressure of 7 mm Hg at the venous end. The plasma oncotic pressure is 15 mm Hg. What transport mechanism causes water to move out of the capillary into the interstitium?
    a. active transport
    b. diffusion
    c. osmosis
    d. filtration

21. The pressures within a capillary are as follows: capillary hydrostatic pressure of 35 mm Hg at the arterial end of the capillary and a pressure of 7 mm Hg at the venous end. The plasma oncotic pressure is 15 mm Hg. What transport mechanism causes water to diffuse from the interstitium into the venous end of the capillary.
    a. active transport
    b. filtration
    c. facilitated diffusion
    d. osmosis

22. Which group is incorrect?
    a. passive transport mechanisms: osmosis, diffusion, Na/K pump
    b. organelles: mitochondria, lysosomes, ribosomes
    c. transport mechanisms: diffusion, osmosis, filtration
    d. stages of mitosis: prophase, anaphase, metaphase, telophase

23. Which group is incorrect?
    a. organelles: mitochondria, lysosomes, ribosomes
    b. types of ribosomes: free, bound (fixed)
    c. concentrations: isotonic, hypertonic, hypotonic
    d. active transport mechanisms: facilitated diffusion, osmosis, pinocytosis

CASE STUDY

While performing breast self-examination, J.S. discovered a lump in the right upper quadrant of her left breast. She immediately contacted her physician, and a biopsy was scheduled. The pathology report indicated a benign neoplasm.

1. Which of the following is most characteristic of a benign neoplasm?
   a. metastatic
   b. well-differentiated cells
   c. malignant
   d. crablike

2. Which of the following words refers to the lump?
   a. biopsy
   b. metastasis
   c. Pap smear
   d. neoplasm

PUZZLE

Hint: How to Get a Cell Mate

Directions: Perform the following functions on the Sequence of Words below. When all the functions have been performed, you are left with a word or words related to the hint. Record your answer below.

Functions: Remove the following:

1. Control center of the cell

2. Hairlike structures (two) on the surface of the cell membrane

3. Passive transport mechanisms (four)

4. Types of endoplasmic reticulum (two)

5. "Power plants" of the cell

6. Organelle that puts the finishing touches on a protein and then packages the protein for export

7. Cellular eating and cellular drinking

8. Phases of mitosis (four)

9. Process of cellular specialization

Sequence of Words

ROUGHMETAPHASEDIFFERENTIATIONOS
MOSISCILIAGOLGIAPPARATUSANAPHASEPI
NOCYTOSISFLAGELLUMNUCLEUSMITO
CHONDRIATELOPHASEMITOSISFACILI
TATEDDIFFUSIONDIFFUSIONPHAGOCYTO
SISPROPHASEFILTRATIONSMOOTH

Answer: _____

4 Cell Metabolism

OBJECTIVES

1. Define *metabolism, anabolism,* and *catabolism.*
2. Explain the use of carbohydrates in the body, and differentiate between the anaerobic and aerobic metabolism of carbohydrates.
3. Explain the use of fats in the body.
4. Explain the use of proteins in the body.
5. Describe the roles of DNA and RNA in protein synthesis, the structure of a nucleotide, and the steps in protein synthesis.

Part I: Mastering the Basics

MATCHING

Carbohydrates, Fats, and Proteins

Directions: Match the following terms to the most appropriate definition by writing the correct letter in the space provided. Some terms may be used more than once. See text, pp. 48-56.

A. glucose
B. amino acids
C. nonessential amino acids
D. disaccharides
E. cellulose
F. glycogen
G. urea
H. nitrogen
I. essential amino acids
J. monosaccharides
K. lipids
L. organic
M. fatty acids and glycerol

1. _____ Refers to carbon-containing substances

2. _____ A nitrogen-containing waste product

3. _____ Building blocks of lipids

4. _____ Nondigestible polysaccharide found in plants

5. _____ Protein contains this in addition to carbon, oxygen, and hydrogen

6. _____ Building blocks joined together by peptide bonds

7. _____ Amino acids that cannot be synthesized by the body and must therefore be obtained through dietary intake

8. _____ Monosaccharide that provides the primary source of energy for cells

9. _____ Building blocks of protein

10. _____ Sucrose, maltose, and lactose; sometimes called *double sugars*

11. _____ Glucose, fructose, and galactose

12. _____ Glucose is stored as this polysaccharide; also called *animal starch*

13. _____ Amino acids that can be synthesized by the body

14. _____ Classification of triglycerides and steroids

15. _____ "Good" cholesterol and "bad" cholesterol

MATCHING

Metabolism of Carbohydrates, Proteins, and Fats

Directions: Match the following terms to the most appropriate definition by writing the correct letter in the space provided. Some terms may be used more than once. See text, pp. 48-56.

A. gluconeogenesis
B. catabolism
C. CO_2, water, energy (ATP)
D. anabolism
E. enzyme
F. Krebs cycle
G. peptide bond
H. lactic acid
I. ketone bodies
J. glycolysis

1. _____ Chemical reactions that build larger, more complex substances

2. _____ Amine group of alanine joins with the acid part of valine to form this.

3. _____ Produced by the rapid incomplete breakdown of fatty acids

4. _____ Chemical reactions that degrade larger, more complex substances into simpler substances

5. _____ Series of reactions that anaerobically break down glucose to lactic acid

6. _____ Series of aerobic reactions that occur in the mitochondria

7. _____ In the absence of oxygen, pyruvic acid is converted to this substance.

8. _____ End products of the aerobic catabolism of glucose

9. _____ Almost every chemical reaction in the body is catalyzed by this.

10. _____ This series of anaerobic reactions occurs within the cytoplasm

11. _____ Conversion of protein into glucose

READ THE DIAGRAM

Directions: Refer to Fig. 4.3 in the textbook, and indicate whether the following statements refer to column A or column B. Write A or B in the blanks below. See text, p. 51.

1. _____ Describes the anaerobic breakdown of glucose to lactic acid

2. _____ Some of the chemical reactions occur within the mitochondria.

3. _____ Describes the complete breakdown of glucose into carbon dioxide, water, and energy

4. _____ Most of the energy is formed in this pathway.

5. _____ Pyruvic acid products enter the Krebs cycle for further catabolism.

6. _____ All chemical reactions occur within the cytoplasm.

7. _____ Illustrates the role of the Krebs cycle in the catabolism of glucose

8. _____ Includes oxygen-requiring reactions

9. _____ Pyruvic acid is not converted to lactic acid.

10. _____ Chemical reactions that catabolize glucose incompletely

MATCHING

Nucleotides: DNA and RNA

Directions: Match the following terms to the most appropriate definition by writing the correct letter in the space provided. Some terms may be used more than once. See text, pp. 56-60.

A. DNA
B. tRNA
C. deoxyribose
D. ribose
E. mRNA
F. nucleotide

1. _____ Double-stranded nucleic acid that contains the genetic code; called the *double helix*

2. _____ Substance composed of phosphate, sugar, and base

3. _____ Sugar found in RNA

4. _____ Sugar found in DNA

5. _____ Nucleotide that copies the genetic code from DNA in the nucleus

6. _____ Nucleotide that carries individual amino acids from the cytoplasm to the ribosomes for assembly along the mRNA

7. _____ DNA and this nucleotide are involved in transcription.

8. _____ mRNA and this nucleotide are involved in translation.

9. _____ Nucleotide that is confined to the nucleus

10. _____ Nucleotide that does not use uracil in its coding; uses thymine instead

11. _____ Nucleotide that carries genetic information from the nucleus to the ribosomes

SIMILARS AND DISSIMILARS

Directions: Circle the word in each group that is least similar to the others. Indicate the similarity of the three words on the line below each question.

1. monosaccharide   polysaccharide
   polypeptide   disaccharide

_____

2. glucose   sucrose   galactose   fructose

_____

3. sucrose   glucose   maltose   lactose

_____

4. amino acids   polypeptide   glycogen   protein

_____

5. metabolism   differentiation   anabolism
   catabolism

_____

6. phenylalanine   starch   glycogen   cellulose

_____

7. cholesterol   phospholipid   glycogen   steroid

_____

8. alanine   leucine   steroid   tyrosine

_____

9. peptide bond   fatty acid $-NH_2$ $-COOH$

_____

10. DNA   tRNA   glycogen   mRNA

_____

11. uracil   thymine   glycogen   adenine

_____

12. cellulose   polysaccharide   starch   urea

_____

13. cellulose   glucose   glycogen   cholesterol

_____

14. peptide bonds   fats   phospholipids   oils

_____

15. phospholipid   protein   steroid   triglyceride

_____

16. essential AA   peptide bonds

     nonessential AA   cholesterol

_____

17. ammonia NH_3 cholesterol   urea

_____

18. G-C   A-T   A-U   G-U

_____

19. T-A   C-G   G-G   A-T

_____

20. tRNA   transcription   gluconeogenesis   mRNA

_____

21. purine   sucrose   adenine   pyrimidine

_____

Part II: Putting It All Together

MULTIPLE CHOICE

Directions: Choose the correct answer.

1. Glucose
   a. is usually burned as fuel to get energy.
   b. can be converted to protein.
   c. cannot be converted to fat.
   d. must be converted to urea to be excreted from the body.

2. Urea is
   a. produced by the kidneys and excreted by the liver into the bile.
   b. a disaccharide.
   c. a protein.
   d. a nitrogen-containing waste produced in the liver and excreted in the urine.

3. Cellulose is
   a. an animal starch.
   b. the storage form of glucose.
   c. a nitrogen-containing waste produced in the liver and excreted in the urine.
   d. a nondigestible carbohydrate.

4. Which of the following best describes the composition of hormones, hemoglobin, and gamma globulins?
   a. carbohydrates
   b. proteins
   c. lipids
   d. glycerol

5. Gluconeogenesis is the process whereby
   a. glucose is converted to protein
   b. fatty acids are combined with glycerol to form a monosaccharide.
   c. protein is used to make glucose.
   d. glucose is broken down to lactic acid.

6. Which of the following is required by the chemical reactions that occur within the mitochondria?
   a. lactic acid
   b. carbon dioxide
   c. oxygen
   d. cellulose

7. Glycogen
   a. combines with three fatty acids to make fat.
   b. is a source of glucose stored in the liver and skeletal muscle.
   c. is synthesized only in the pancreas.
   d. is a monosaccharide.

21

8. Glucose, fructose, and galactose
   a. form peptide bonds.
   b. are lipids.
   c. can only be metabolized aerobically.
   d. are monosaccharides.

9. When the blood sugar level decreases, the glycogen in the liver is converted into which substance?
   a. protein
   b. ATP
   c. glucose
   d. lactic acid

10. Glycolysis
    a. is aerobic.
    b. forms CO_2 + water + energy.
    c. occurs only within the mitochondria.
    d. is anaerobic and cytoplasmic.

11. Which of the following is most related to the reactions of the Krebs cycle?
    a. anaerobic
    b. mitochondrial
    c. lactic acidosis
    d. urea-forming

12. Which of the following conditions is caused by a lack of oxygen in a critically ill patient?
    a. lactic acidosis
    b. cancer
    c. decreased blood glucose level
    d. hypertension

13. Which of the following is least descriptive of ammonia?
    a. toxic to the brain
    b. accumulates in the presence of liver failure
    c. nitrogen-containing
    d. primary fuel for "running" the body

14. This protein substance acts as a catalyst, increasing the rate of a chemical reaction.
    a. urea
    b. enzyme
    c. mRNA
    d. ribose

15. If the bases of one side of DNA read A-G-C-T, the complementary (opposite) DNA strand reads
    a. A-C-G-T.
    b. A-G-A-T.
    c. U-C-G-A.
    d. T-C-G-A.

16. Which of the following is most related to the storage of the genetic code?
    a. Krebs cycle
    b. glycolysis
    c. mitosis
    d. base-sequencing within the DNA molecule

17. With regard to base pairing, thymine can only pair with which base?
    a. adenine
    b. uracil
    c. cytosine
    d. guanine

18. The bases in a strand of DNA read T-C-C-A. The transcribed strand of mRNA reads
    a. A-G-G-T.
    b. G-C-G-A.
    c. U-C-C-T.
    d. A-G-G-U.

19. The double helix refers to
    a. the growing peptide chain sitting along the ribosome
    b. the tRNA attached to amino acids in the cytosol
    c. the double-stranded mRNA involved in transcription
    d. the double-stranded DNA in the nucleus

20. What is described: cytoplasmic, single-stranded nucleotide, and uracil-containing?
    a. essential amino acids
    b. DNA
    c. RNA
    d. albumin

21. Transcription
    a. occurs within the nucleus.
    b. refers to the interaction between mRNA and tRNA.
    c. refers to the synthesis of a complimentary strand of DNA.
    d. refers to the assembly of amino acids along the ribosomes.

22. As a person ages, the body is less able to metabolize glucose efficiently, which results in elevated
    a. blood calcium levels.
    b. blood pressure.
    c. serum cholesterol levels.
    d. blood sugar levels.

23. A hormone or drug that suppresses gluconeogenesis
    a. prevents the elevation of the blood glucose level.
    b. produces ketone bodies.
    c. prevents the synthesis of urea by the liver.
    d. prevents the secretion of urea by the kidneys.

24. Which group is incorrect?
    a. monosaccharides: glucose, fructose, galactose
    b. nucleotides: DNA, RNA
    c. lipid-related structures: cholesterol, steroid, fatty acids
    d. amino acids: sucrose, maltose, lactose

25. Which group is incorrect?
    a. amino acids: essential, nonessential
    b. monosaccharides: glucose, fructose, galactose
    c. fatty acids: glyccrol, glycogen
    d. ketone bodies: keto acids, acetone

CASE STUDY

Until recently, 6-year-old Billie had no apparent health problems. About 1 week ago, she started to lose weight despite a healthy appetite. She urinated frequently and complained of being tired. Her mom noticed that she was very thirsty and was getting up in the middle of the night to urinate. On cxamination, her blood sugar level was elevated, and she had sugar and acetone in her urine. She was diagnosed with type 1 (juvenile-onset) diabetes mellitus.

1. Because the diabetic cells cannot use glucosc, it accumulates in the blood, causing this condition.
    a. hypotension
    b. hyperglycemia
    c. lactic acidosis
    d. alkalosis

2. Because the diabetic person cannot use glucose, Billie metabolizes fatty acids rapidly and incompletely, thereby producing excessive amounts of this (these) substance(s).
    a. enzymes
    b. kctone bodies
    c. glycogen
    d. urea

3. Which of the following describes the effect of excess ketone bodies in the blood of a diabetic person?
    a. glucosuria
    b. acidosis
    c. elevated blood pH
    d. hyperglycemia

PUZZLE

Hint: Needs Oxygen . . . Name and Address

Directions: Perform the following functions on the Sequence of Words that follows. When all the functions have been performed, you are left with a word or words related to the hint. Record your answer in the space provided.

Functions: Remove thc following:

1. Building blocks of proteins

2. Monosaccharides

3. Nitrogen-containing waste product

4. Series of reactions that degrade glucose to lactic acid; anaerobic and cytoplasmic

5. Series of reactions that make glucose from protein breakdown products

6. Disaccharides

7. Building blocks of fats

8. Storage form of glucose

9. Term that includes anabolism and catabolism

10. Nucleic acid that stores the genetic code within the nucleus

11. Nucleic acid that carries the genetic code from the nucleus to the ribosomes in the cytoplasm

12. Two sugars found in DNA and RNA

13. Four bases found in DNA

14. Type of bond that forms between amino acids

15. The core that forms VLDL, LDL, and HDL

16. NH_3

Sequence of Words

GUANINEAMMONIARIBOSEDNAGLUCONE
OGENESISSUCROSEGLUCOSECYTOSIN
EMALTOSEGALACTOSEUREAFATTYACI
DSPEPTIDEBONDGLYCOGENGLYCEROLM
ETABOLISMDEOXYRIBOSEAMINOACI
DSCHOLESTEROLTHYMINEHANSKREBSM
ITOCHONDRIONADENINEGLYCOLYSISFRU
CTOSEmRNALACTOSE

Answer: _____

5 Microbiology Basics

OBJECTIVES

1. Define *disease* and *infection*.
2. List the characteristics of the different types of pathogens, including the types of bacteria by shape.
3. Describe the types of bacteria by staining characteristics.
4. Define portals of exit and portals of entry.
5. List common ways by which infections are spread.
6. Identify the microbiological principles described in Five Germ-Laden Stories.

Part I: Mastering the Basics

MATCHING

Germs, Worms, and Terms

Directions: Match the following terms to the most appropriate definition by writing the correct letter in the space provided. See text, pp. 65-69.

A. parasites	G. disease
B. vector	H. portals of entry
C. normal flora	I. pathogen
D. Gram stain	J. infection
E. zoonosis	K. portals of exit
F. nosocomial	

1. _____ Failure of the body to function normally

2. _____ Disease-causing organism

3. _____ Disease caused by a pathogen or its toxin

4. _____ An object (living or nonliving) that transfers a pathogen from one organism to another

5. _____ Organisms that normally and harmoniously live in or on the human body; also called microbiota

6. _____ Organisms that require a living host in which to survive

7. _____ Routes whereby pathogens enter the body

8. _____ Hospital-acquired infection

9. _____ Dye used to identify different types of bacteria

10. _____ An animal disease transmissible to humans

11. _____ Routes whereby pathogens leave the body

MATCHING

Pathogens

Directions: Match the following terms to the most appropriate definition by writing the correct letter in the space provided. Some terms may be used more than once. See text, pp. 65-69.

A. bacteria	D. protozoa
B. virus	E. worms
C. fungi	F. arthropods

1. _____ Coccus, bacillus, curved rod

2. _____ Acts as a parasite to the infected cell

3. _____ Mycotic infections

4. _____ Plantlike organisms such as mushrooms

5. _____ Single-cell, animal-like microbes

6. _____ *Chlamydia* and *Rickettsia*

7. _____ From the Latin meaning *poison*

8. _____ Amebas, ciliates, flagellates, sporozoa

9. _____ *Diplococcus, Streptococcus, Staphylococcus*

10. _____ Consists of RNA or DNA surrounded by a protein shell

11. _____ Yeasts and molds

12. _____ Arrangement: pairs, chains, bunches of grapes

13. _____ *Vibrio, Spirillum*, spirochete

14. _____ Helminths

15. _____ Ectoparasites

16. _____ *Ascaris, Trichina*, flukes

17. _____ Animals with jointed legs, including insects and ticks

18. _____ Gram-positive (+) and gram-negative (−)

MATCHING

Five Germ-Laden Stories

Directions: Match the following stories to the most appropriate description by writing the correct letter in the space provided. The stories may be used more than once. See text, pp. 71-76.

A. Dr. Semmelweis Screams "Wash Those Mitts"
B. Flora and Her Vaginal Itch
C. Rick, Nick, and the Sick Tick
D. Why Typhoid Mary Needed to Lose Her Gallbladder
E. Pox News Alert

1. _____ Focuses on handwashing, dirty hands, and nosocomial infection

2. _____ Great example of the carrier state and disease transmission by the fecal-oral route

3. _____ Focuses on changes in microbiota and superinfection

4. _____ Varicella-zoster

5. _____ *Salmonella* hides in bile.

6. _____ The injected saliva was teeming with rickettsiae.

7. _____ The conclusion of this tormented scientist: "Puerperal fever is caused by conveyance to the pregnant woman of 'putrid particles' derived from living organisms through the agency of the examining finger."

8. _____ Great example of zoonosis

9. _____ Describes an insect as a "reservoir of infection"

10. _____ Describes the overgrowth of *Candida albicans* (yeast), causing itching and discharge

11. _____ Vesicular skin lesions

12. _____ Describes the alteration of the normal flora by antibiotics

13. _____ This streptococcal infection began in the uterus and progressed to peritonitis, generalized septicemia, and death.

14. _____ Illustrates disease transmission by an arthropod vector

15. _____ Childhood illness and shingles

16. _____ Differentiates between a communicable and contagious disease

17. _____ *Salmonella*-laced dinner

SIMILARS AND DISSIMILARS

Directions: Circle the word in each group that is least similar to the others. Indicate the similarity of the three words on the line below each question.

1. curved rods   viruses   cocci   bacilli
   _____

2. fungus   coccus   athlete's foot   mycotic infection
   _____

3. *Vibrio*   *Spirillum*   fungi   spirochetes
   _____

4. amebas   ciliates   spirochetes   flagellates
   _____

5. ticks   fleas   mosquitoes   bacilli
   _____

6. gram-positive (+)   culture   gram-negative (−)   acid-fast
   _____

7. myco-   diplo-   staphylo-   strepto-
   _____

8. yeast   virus   mold   fungus
   _____

9. *Clostridium*   anaerobic   typhoid fever   tetanus
   _____

10. nosocomial   hospital-acquired   infection   vaccine
    _____

11. animal vector   mosquito   arthropod vector   streptococcus
    _____

12. luetic   syphilitic   malaria   spirochete
    _____

13. fomite   nonliving   contaminated   mosquito vector   syringe

_____

14. herpes   food   shingles   varicella-zoster zoster   poisoning

_____

15. tetanus   Typhoid Mary   fecal-oral   carrier route

_____

16. *Salmonella*   anaerobic   fecal-oral   typhoid fever route

_____

Part II: Putting It All Together

MULTIPLE CHOICE

Directions: Choose the correct answer.

1. Ringworm
   a. refers to a helminthic infection.
   b. is a mycotic infection.
   c. is also called impetigo.
   d. is caused by a strain of staphylococcus.

2. Person-to-person, environment-to-person, and tiny animal-to-person refer to the
   a. staining characteristics of pathogens.
   b. size of the pathogens.
   c. spread or transmission of the pathogens.
   d. sensitivity to chemotherapy.

3. Which of the following is a true statement about bacteria?
   a. All bacteria are pathogenic.
   b. The spread of bacterial infections requires arthropod vectors.
   c. Pathogenic bacteria cause mycotic infections.
   d. Most bacteria perform useful functions; fewer are pathogenic.

4. Mycotic infections are
   a. caused by gram-negative (–) bacilli.
   b. viral infections.
   c. fungal infections.
   d. always zoonotic.

5. Infection by *Candida albicans*
   a. is viral and therefore unresponsive to antibiotics.
   b. only occurs in the vagina.
   c. often appears as a superinfection after antibiotic therapy.
   d. requires a biological vector such as a tick.

6. Viruses
   a. are surrounded by a rigid cell wall.
   b. are sensitive to penicillin.
   c. are the smallest of the infectious agents and are fragments of either RNA or DNA surrounded by a protein shell.
   d. are classified as gram-positive (+) and gram-negative (–).

7. Amebas, flagellates, ciliates, and sporozoa are
   a. anaerobic, spore-forming bacteria.
   b. protozoa.
   c. fungi.
   d. rickettsiae.

8. An anthelmintic exerts its therapeutic effects against
   a. gram-negative (–) bacteria.
   b. all spore-forming microorganisms.
   c. viruses.
   d. worms.

9. A young child has pinworms. Which observation will usually be made by the parent?
   a. malaise, fever, and generalized wasting
   b. abdominal swelling and severe diarrhea
   c. nausea, vomiting, and weight loss
   d. perianal pruritus (itching)

10. These arthropods live on the surface of the body and cause itching and discomfort but are not life-threatening.
   a. ectoparasites
   b. normal flora
   c. viruses
   d. protozoa

11. Regarding the spread of malaria, the mosquito is the
   a. causative organism of malaria.
   b. arthropod vector.
   c. fomite.
   d. pathogen.

12. Which of the following is least related to staphylococcus?
   a. gram-positive (+) bacterium
   b. impetigo, food poisoning, and scalded skin syndrome
   c. chickenpox and shingles
   d. usual cause of most skin infections

13. *Clostridium tetani*
   a. is an ectoparasite.
   b. grows best in a puncture wound under anaerobic conditions.
   c. is gram-positive (+) and aerobic.
   d. grows best on the skin surface.

14. Which of the following is not characteristic of *Candida albicans*?
    a. mycotic infection
    b. member of the normal flora of the mouth, digestive tract, and vagina
    c. spore-forming anaerobe like the *Clostridium* microbes
    d. can cause a superinfection

15. A pox is
    a. usually responsive to an antibiotic cream.
    b. a mycotic infection.
    c. a vesicular skin lesion.
    d. a luetic lesion.

16. The Great Pox, chancre, and luetic lesions are associated with
    a. gonorrhea.
    b. giardiasis.
    c. impetigo.
    d. syphilis.

17. Which group is incorrect?
    a. arrangement of cocci: diplococcus, streptococcus, staphylococcus
    b. bacteria: coccus, bacillus, curved rod
    c. curved rods: *Vibrio, Spirillum*, fungus
    d. arthropod vectors: fleas, ticks, lice, mosquitoes

18. Which group is incorrect?
    a. bacteria: coccus, bacillus, curved rod
    b. arthropod vectors: fleas, ticks, lice, mosquitoes
    c. protozoa: amebas, ciliates, flagellates, sporozoa
    d. contagious diseases: measles, chickenpox, urinary bladder infection

19. Which group is incorrect?
    a. arrangement of cocci: diplococcus, streptococcus, bacillus
    b. arthropod vectors: fleas, ticks, lice, mosquitoes
    c. protozoa: amebas, ciliates, flagellates, sporozoa
    d. contagious diseases: measles, chickenpox, mumps

PUZZLE

Hint: Fun Guy or Fungi

Directions: Perform the following functions on the Sequence of Words that follows. When all the functions have been performed, you are left with a word or words related to the hint. Record your answer in the space provided.

Functions: Remove the following:

1. Types of bacteria

2. Hospital-acquired infection

3. Disease-causing microorganism

4. Organisms that normally and harmoniously live in or on the human body without causing disease (2 terms)

5. Two types of cocci

6. Organism that requires a living host

7. A parasitic worm

8. A tiny parasite that is composed primarily of strands of DNA or RNA

9. Refers to mites, lice, and ticks

10. An animal disease that is transmissible to humans

11. Main types (four) of protozoa

12. This disease is caused by the acid-fast bacillus.

13. The most famous spirochete (*Treponema pallidum*) causes this luetic infection.

14. To what disease do these terms refer: "clap," STD, and *Neisseria*?

15. The microorganism that causes a deadly form of food poisoning

16. The animal vector that carries the causative organism of malaria.

Sequence of Words

AMEBASSYPHILLISCOCCUSPARASITENOS
OCOMIALCILIATESSTAPHECTOPARASITES
CLOSTRIDIUMBOLTULINUMFLAGEL
LATESZOONOSISTUBERCULOSISNORMAL
FLORAVIRUSSPOROZOAMOSQUITOBACIL
LUSHELMINTHMICROBIOTAMOLDCURVE
DRODSGONORRHEASTREPPATHOGEN

Answer: _____

6 Tissues and Membranes

OBJECTIVES

1. List the four major types of tissues.
2. Do the following regarding epithelial tissue:
   - Describe the characteristics and functions of epithelial tissue.
   - Explain how epithelial tissue is classified.
   - List the types of epithelial tissue membranes.
   - Differentiate between endocrine and exocrine glands.
3. Describe the characteristics and functions of connective tissue, and list the types of connective tissue membranes.
4. Describe the characteristics and functions of nervous and muscle tissues.
5. Explain the process of tissue repair after an injury.
6. Differentiate between mucous and serous membranes.

Part I: Mastering the Basics

MATCHING

Types of Tissues

Directions: Match the following terms to the most appropriate definition by writing the correct letter in the space provided. Some terms may be used more than once. See text, pp. 77-88.

A. epithelial tissue   C. muscle tissue
B. connective tissue   D. nervous tissue

1. _____ Tissue that is avascular and is nourished from the underlying connective tissue

2. _____ Osseous tissue

3. _____ Attached to a basement membrane

4. _____ Blood, bone, cartilage, and adipose tissue

5. _____ Neurons and glia

6. _____ Classified as squamous, cuboidal, or columnar

7. _____ Classified as simple or stratified

8. _____ Has the most diverse and greatest amount of intercellular matrix of the four tissue types

9. _____ Classified as skeletal, smooth, or cardiac

10. _____ Type of tissue that forms tough bands that attach muscle to bone

11. _____ Dense fibrous, reticular, and areolar

12. _____ Specialized type of this tissue stores fat

13. _____ A sarcoma arises from this type of tissue.

14. _____ A carcinoma arises from this type of tissue.

15. _____ Primary functions include secretion, absorption, excretion, and protection.

16. _____ Most abundant of the four tissue types

17. _____ Forms the epidermis

18. _____ Endocrine and exocrine glands arise from this type of tissue.

19. _____ Binds together parts of the body; examples include ligaments, tendons, capsules, and fascia

20. _____ Has two surfaces; one surface is always unattached or free, such as the surface of the outer skin and the lining of the mouth

21. _____ Chondrocytes and osteocytes

22. _____ Transitional; found in stretchy organs such as the urinary bladder

MATCHING

Membranes

Directions: Match the following terms to the most appropriate definition by writing the correct letter in the space provided. See text, pp. 88-89.

A. visceral pleura   F. parietal peritoneum
B. synovial membrane   G. cutaneous membrane
C. visceral peritoneum   H. mucous membrane
D. parietal pleura   I. connective tissue
E. pericardium   membrane

1. _____ Membrane lining all body cavities that open to the outside of the body

2. _____ Connective tissue membrane that lines the cavities of joints

3. _____ Skin

4. _____ Serous membrane that covers the outside of each lung

5. _____ Serous membrane that lines the inner wall of the abdominopelvic cavity

6. _____ Serous membrane that lines the organs of the abdominopelvic cavity

7. _____ Serous membrane that lines the walls of the thoracic cavity

8. _____ Synovial membrane and periosteum

9. _____ Lines the mouth, nose, and respiratory passages

10. _____ Sling that supports the heart

READ THE DIAGRAM

Directions: Referring to the illustration (Fig. 6.9 in the text), write the numbers on the lines provided below. See text, pp. 88-89.

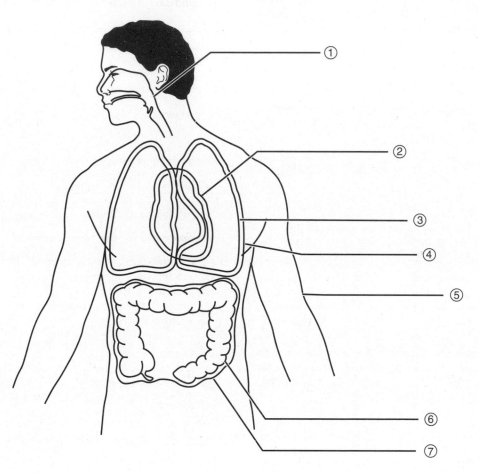

1. _____ Example of a mucous membrane

2. _____ Cutaneous membrane

3. _____ Parietal peritoneum

4. _____ Serous membrane that "hugs" the outer surface of the lung

5. _____ A pericardial membrane

6. _____ Visceral peritoneum

7. _____ Serous membrane that "hugs" the inner walls of the thoracic cavity

8. _____ Membrane that surrounds the heart

9. _____ The intrapleural space is located between 4 and _____

10. _____ Most likely to become congested with the common cold

COLOR AND DRAW

Directions: Using the diagram from p. 30, color the following structures.

1. Color the parietal pleura *blue*.
2. Color the visceral pleura *green*.
3. Color the intrapleural space *yellow*.
4. Draw in the diaphragm.

SIMILARS AND DISSIMILARS

Directions: Circle the word in each group that is least similar to the others. Indicate the similarity of the three words on the line below each question.

1. epithelial   pleura   muscle   connective

   _____

2. areolar   cuboidal   squamous   columnar

   _____

3. exocrine   glandular   meninges   endocrine
   epithelium

   _____

4. collagen   reticular fibers   simple   elastin
                                      squamous

   _____

5. fish scale–like   tendons   dice-   column-
                                    shaped   shaped

   _____

6. areolar   dense fibrous   glandular   osseous
                                  epithelium

   _____

7. osseous   glandular   blood   loose connective
               epithelium

   _____

8. hyaline   adipose   fibrocartilage   elastic

   _____

9. fat   pleura   adipose   connective

   _____

10. cuboidal   tendons   fascia   ligaments

   _____

11. skeletal   glia   smooth   cardiac

   _____

12. cartilaginous   connective   nervous   osseous

   _____

13. mucous   serous   endocrine   cutaneous

   _____

14. synovial   mucous   periosteum   meninges

   _____

15. lipoma   adenoma   peritonitis   osteoma

   _____

16. melanoma   decubitus   bedsore   pressure
                 ulcer   ulcer

   _____

Part II: Putting It All Together

MULTIPLE CHOICE

Directions: Choose the correct answer.

1. The parietal and visceral pleurae
   a. are mucous membranes.
   b. secrete small amounts of serous fluid.
   c. are located in the abdominal cavity.
   d. surround the heart.

2. Epithelial tissue
   a. has extensive intercellular material.
   b. forms large continuous sheets of tissue.
   c. forms tendons, ligaments, and capsules.
   d. is described as visceral and parietal.

3. Glandular tissue
   a. is found only within the abdominal organs.
   b. arises from epithelial tissue.
   c. is classified as dense, fibrous, and areolar.
   d. stores fat.

4. Mucous membrane
   a. forms the pleurae.
   b. forms the peritoneal membranes.
   c. lines the respiratory tract.
   d. is a type of connective tissue membrane.

5. Squamous, cuboidal, and columnar
   a. refer to the layers of epithelial tissue.
   b. are types of nervous tissues.
   c. are shapes of epithelial tissues.
   d. are found only within the thoracic cavity.

6. Simple and stratified
   a. refer to the layers of epithelial tissue.
   b. are types of nervous tissues.
   c. are shapes of epithelial tissues.
   d. are found only within the thoracic cavity.

7. Which of the following membranes is confined to the thoracic cavity?
   a. meninges
   b. peritoneum
   c. synovial membranes
   d. pleurae

8. Because this type of tissue is so thin, it is concerned primarily with the movement of various substances across the membranes from one body compartment to another.
   a. connective tissue
   b. neuroglia
   c. simple squamous epithelium
   d. cartilage

9. Which of the following is most related to glandular epithelium?
   a. cuboidal epithelium
   b. skeletal, cardiac, and smooth
   c. tendons, ligaments
   d. adipose

10. Which of the following are related to endocrine glands?
    a. pleura and peritoneum
    b. serous and mucous
    c. ductless glands
    d. visceral and parietal

11. In which type of tissue is the intercellular matrix hardest?
    a. blood
    b. osseous
    c. simple squamous epithelium
    d. adipose

12. Which of the following is most descriptive of cartilage?
    a. squamous cell epithelium
    b. basement membrane
    c. hyaline and elastic
    d. endocrine and exocrine

13. Which of the following does not appear in the thoracic cavity?
    a. serous membranes
    b. pleural membranes
    c. serous fluid
    d. peritoneal membrane

14. Which of the following best describes scar tissue?
    a. sarcoma
    b. regeneration
    c. meninges
    d. fibrosis

15. Which of the following are described as parietal and visceral?
    a. mucous membranes
    b. serous membranes
    c. synovial membranes
    d. meninges

16. This condition is caused by prolonged pressure that results in a decrease in the blood supply to the tissues.
    a. peritonitis
    b. lactic acidosis
    c. decubitus ulcer
    d. meningitis

17. Which word is most descriptive of gangrenous tissue?
    a. necrotic
    b. scar
    c. fibrotic
    d. malignant

18. Why do tissues become stiffer and less efficient with aging?
    a. The epithelial membranes become thicker.
    b. There is an increase in intracellular fluid.
    c. There is a decrease in collagen and elastin in connective tissue.
    d. Muscle and nerve tissue hypertrophy.

19. Which group is incorrect?
    a. appearance of epithelial tissue: squamous, cuboidal, columnar
    b. types of muscles: skeletal, cardiac, and adipose
    c. types of nervous tissues: neurons and neuroglia
    d. types of fibers: collagen, elastin, and reticular

20. Which group is incorrect?
    a. shapes of epithelial tissues: squamous, cuboidal, and columnar
    b. types of connective tissues: adipose, areolar, and dense fibrous
    c. layers of epithelial tissue: simple and stratified
    d. serous membranes in thoracic cavity: pleura, pericardium, and peritoneum

21. Which group is incorrect?
    a. layers of connective tissue: simple and stratified
    b. types of tissues: epithelial, connective, nerve, and muscle
    c. types of nervous tissues: neurons and neuroglia
    d. types of connective tissues: areolar, dense fibrous, reticular, cartilage, bone, and blood

CASE STUDY

Six-year-old J.P. complained to her mom that she had a pain in her stomach. Her mom attributed the pain to too much junk food and suggested that she go to bed early. At 4 AM, J.P. awoke with severe abdominal pain, a high fever, and a rigid boardlike abdomen. Her physician admitted her to the hospital with a diagnosis of a ruptured appendix.

1. Which of the following is likely to develop in response to the ruptured appendix as waste (feces) leaks into the abdominopelvic cavity?
   a. peritonitis
   b. pleurisy
   c. hemorrhoids
   d. pericarditis

2. Which of the following became inflamed?
   a. pleura
   b. serous membrane
   c. meninges
   d. synovial membrane

PUZZLE

Hint: Scales, Columns, and Cubes

Directions: Perform the following functions on the Sequence of Words below. When all the functions have been performed, you are left with a word or words related to the hint. Record your answer below.

Functions: Remove the following:

1. Examples of connective tissues (six)

2. Two types of nervous tissues

3. Types of muscle tissues (three)

4. Two types of glands formed from glandular epithelium

5. Three serous membranes

6. Types of connective tissues that have a watery intercellular matrix

7. Type of membrane that lines the digestive and respiratory tracts

8. Connective tissue membrane that covers bone

9. Connective tissue membrane that covers the brain and spinal cord

Sequence of Words

MENINGESCARTILAGEEXOCRINESMOOTH
PERITONEUMTENDONSNEUROGLIAAREO
LARSKELETALPERIOSTEUMBONEPERICAR
DIUMENDOCRINEBLOODEPITHELIALTIS
SUEADIPOSELYMPHPLEURALIGAMENTSCA
RDIACMUCOUSNEURON

Answer: _____

7 Integumentary System and Body Temperature

OBJECTIVES

1. List seven functions of the skin.
2. Discuss the structure of the skin, including these actions:
   - Describe the two layers of skin: the epidermis and the dermis.
   - Define *stratum germinativum* and *stratum corneum*.
   - List the two major functions of the subcutaneous layer.
3. List the factors that influence the color of the skin.
4. Describe the accessory structures of the skin: hair, nails, and glands.
5. Define *thermoregulation*, and describe the way that the body conserves and loses heat.
6. Differentiate between insensible and sensible perspiration.

Part I: Mastering the Basics

MATCHING

The Skin

Directions: Match the following terms to the most appropriate definition by writing the correct letter in the space provided. Some terms may be used more than once. See text, pp. 93-96.

A. dermis
B. subcutaneous layer
C. epidermis
D. keratin
E. stratum germinativum
F. stratum corneum
G. integument
H. dermatology

1. _____ Study of the skin and skin disorders

2. _____ Layer that contains the stratum germinativum and stratum corneum

3. _____ Layer that contains adipose tissue

4. _____ Layer of skin that contains the blood vessels, nerves, and sensory receptors

5. _____ Epidermal layer that forms the bathtub ring

6. _____ Layer underneath the epidermis

7. _____ Layer of epidermis that continuously produces millions of cells every day

8. _____ A protein in the skin that flattens and hardens the skin and makes the skin water-resistant

9. _____ Surface layer of the epidermis that makes up most of the epidermal thickness

10. _____ Layer that insulates the body from extreme temperature changes and anchors the skin to the underlying structures

11. _____ Hypodermis

12. _____ Another name for the skin or cutaneous membrane

13. _____ Desquamation and exfoliation refer to this epidermal layer.

14. _____ This epidermal layer forms corns and calluses.

15. _____ Epidermal layer that sits on the dermis

MATCHING

Glands of the Skin

Directions: Match the following terms to the most appropriate definition by writing the correct letter in the space provided. Some terms may be used more than once. See text, pp. 99-100.

A. ceruminous
B. apocrine
C. eccrine
D. sebaceous
E. mammary

1. _____ Oil glands

2. _____ A blackhead is formed when this gland becomes blocked by accumulated oil and cellular debris.

3. _____ A pimple is formed when the sebum of this blocked gland becomes infected.

4. _____ Sweat glands that cause body odor and are responsible for sex attractants in animals

5. _____ Babies are born with a "cream cheese–like" covering called the *vernix caseosa* that is secreted by these glands.

6. _____ These glands and the apocrine glands are classified as sudoriferous glands.

7. _____ Sweat glands that are usually associated with hair follicles and are found in the axillary and genital areas

8. _____ Sweat glands that respond to emotional stress and become activated when a person is sexually aroused

9. _____ Sweat glands that play the most important role in temperature regulation

10. _____ Milk-secreting glands classified as modified sweat glands

11. _____ Modified sweat glands that secrete ear wax

MATCHING

Skin Characteristics and Conditions

Directions: Match the following terms to the most appropriate definition by writing the correct letter in the space provided. See text, pp. 97-99.

A. jaundice
B. freckles
C. melanoma
D. cyanosis
E. melanin
F. albinism
G. carotene
H. vitiligo
I. alopecia
J. flushing
K. pallor
L. hirsutism

1. _____ Pigment that darkens the skin; pigment-secreting cells stimulated by ultraviolet radiation (i.e., tanning)

2. _____ Condition in which no melanin is secreted

3. _____ Loss of pigment in the skin that creates patches of white skin

4. _____ Melanin is concentrated in local areas; benign

5. _____ Highly malignant form of skin cancer that arises from melanocytes

6. _____ Substance that gives persons of Asian descent a slight yellowish skin coloring

7. _____ Caused by vasodilation of the dermal blood vessels

8. _____ Caused by vasoconstriction of the dermal blood vessels

9. _____ Condition in which the skin has a bluish tint caused by a diminished amount of oxygen in the blood

10. _____ Color change caused by hypoxemia

11. _____ Shaggy or hairy

12. _____ Yellowing of the skin caused by an accumulation of bilirubin in the skin

13. _____ Described as ashen

14. _____ Hair loss

MATCHING

Skin: Water and Temperature

Directions: Match the following terms to the most appropriate definition by writing the correct letter in the space provided. See text, pp. 101-104 .

A. sensible perspiration
B. hypothalamus
C. insensible perspiration
D. thermoregulation
E. shivering thermogenesis
F. modes of heat loss
G. core temperature
H. shell temperature
I. nonshivering thermogenesis
J. pyrexia
K. hyperthermia

1. _____ Temperature within the cranial, thoracic, and abdominal cavities

2. _____ Mechanisms whereby the body balances heat production and heat loss

3. _____ About 500 mL/day of water is lost through the skin and lungs.

4. _____ Eccrine glands are responsible for this type of perspiration during exercise and elevated environmental temperature.

5. _____ Heat production caused by continuous contractions of skeletal muscles when cold

6. _____ Evaporation, conduction, convection, and radiation

7. _____ Heat production caused by the metabolism of brown fat in the neonate

8. _____ Temperature of the surface areas such as the skin, mouth, and axilla

9. _____ Thermostat of the body

10. _____ Another name for fever

11. _____ Elevation of body temperature that is secondary to an increase in the temperature set point

12. _____ Elevation in body temperature in which the body cannot rid itself of excess heat; temperature set point not elevated

13. _____ Elevated temperature that is treated with drugs such as aspirin

14. _____ Type of temperature elevation that is often caused by an infection

15. _____ Type of temperature elevation that is caused by exercising in excessively hot weather

SIMILARS AND DISSIMILARS

Directions: Circle the word in each group that is least similar to the others. Indicate the similarity of the three words on the line below each question.

1. dermis   stratum corneum   subcutaneous   epidermis

_____

2. subcutaneous   hypodermis   adipose tissue   eccrine

_____

3. eccrine   apocrine   sudoriferous   alopecia

_____

4. apocrine   sweat   sebum   eccrine

_____

5. jaundice   moles   freckles   vitiligo

_____

6. jaundice   sweat   cyanosis   pallor

_____

7. yellow   jaundice   cyanosis   hirsutism

_____

8. hypoxemia   blue   albinism   cyanosis

_____

9. melanin   sebum   carotene   bilirubin

_____

10. shaft   lunula   root   follicle

_____

11. sebaceous   brown fat   sebum   vernix caseosa

_____

12. cerumen   modified ear wax   pyrexia   sweat gland

_____

13. evaporation   filtration   conduction   convection

_____

Part II: Putting It All Together

MULTIPLE CHOICE

Directions: Choose the correct answer.

1. Which of the following is least true of the epidermis?
   a. thin outer layer of the skin
   b. contains a rich supply of blood vessels
   c. contains the stratum germinativum and stratum corneum
   d. cells become keratinized

2. Epidermal cells become keratinized, meaning that they
   a. appear yellow.
   b. are hardened and water-resistant.
   c. have been replaced by connective tissue.
   d. appear cyanotic.

3. The dermis
   a. contains the stratum germinativum and stratum corneum.
   b. contains the blood vessels that oxygenate the epidermis.
   c. cannot stretch.
   d. is the target for subcutaneous injections.

4. The subcutaneous layer
   a. is located directly under the epidermis.
   b. contains the stratum germinativum.
   c. is the layer that tans when exposed to ultraviolet radiation.
   d. supports the dermis.

5. Why do we feel the heat intensely on a hot and humid day?
   a. Heat loss increases because of an increase in evaporation.
   b. Vasoconstriction traps heat in the body.
   c. Heat loss decreases because of a decrease in evaporation.
   d. The arrector pili muscles work overtime.

6. Epidermal cells desquamate or exfoliate, meaning that they
   a. become yellow.
   b. form calluses.
   c. slough.
   d. synthesize vitamin D in response to ultraviolet radiation.

7. Which of the following is true of the stratum germinativum?
   a. The cells are dead and are constantly sloughed off.
   b. It is part of the dermis.
   c. It originates in the hypodermis.
   d. It replenishes the cells of the epidermis that have sloughed off.

8. The skin over your knuckles is wrinkled and creased because it is
   a. keratinized.
   b. avascular.
   c. dead.
   d. anchored directly to bone.

9. Albinism, freckles, vitiligo, moles, and tanning are conditions that are all associated with which of the following?
   a. cancer
   b. melanin
   c. cell necrosis
   d. keratinization

10. Alopecia is most likely to occur in a
    a. tanned patient.
    b. cyanotic patient.
    c. patient who is being treated with anticancer drugs.
    d. patient with vitiligo.

11. Which of the following is related to eccrine glands?
    a. classified as endocrine glands
    b. secrete sebum
    c. secrete sweat
    d. called *oil glands*

12. Nonshivering thermogenesis in the neonate is
    a. accomplished by contracting skeletal muscles.
    b. caused by the contraction of the arrector pili muscles.
    c. a heat loss mechanism.
    d. accomplished by the metabolism of brown fat.

13. Which of the following people is most likely to exhibit hirsutism?
    a. person who is chronically hypoxemic (low oxygen) because of a lung disease
    b. person who has liver failure and therefore has excess bilirubin in the blood
    c. woman who is taking testosterone
    d. male who has been castrated (loss of testicular function)

14. Which of the following people is most likely to exhibit clubbing?
    a. person who is chronically hypoxemic (low oxygen) because of a lung disease
    b. person who has liver failure and therefore has excess bilirubin in the blood
    c. woman who is taking testosterone
    d. male who has been castrated (loss of testicular function)

15. Which patient is most likely to require an escharotomy?
    a. person who is cyanotic
    b. person with severe digital clubbing
    c. severely burned patient
    d. person with drug-induced urticaria

16. Why does the skin become dry, coarse, and itchy in the older person? There is
    a. a decrease in the amount of adipose tissue.
    b. an increase in the blood flow to the skin.
    c. an increased fragility of the skin.
    d. decreased sebaceous gland activity in the skin.

17. Why do older people tend to feel cold?
    a. There is a decrease in the amount of adipose tissue under the skin.
    b. There is an increase of blood flow to the skin, resulting in excess cooling.
    c. Metabolic rate increases with age.
    d. Keratinization decreases with age.

18. An infant is left unattended in a car for 2 hours; the temperature in the car was measured as being more than 115°F. Which of the following is a true statement?
    a. The infant will respond successfully with fluid replacement and aspirin.
    b. The infant is described as jaundiced.
    c. The infant is described as hyperthermic.
    d. The infant will survive because of nonshivering thermogenesis.

19. Which of the following best describes urticaria and pruritis?
    a. shingles and pain
    b. hives and itching
    c. acne and redness
    d. callus and numbness

20. A nitroglycerine patch is placed on the skin of a person with a history of chest pain induced by exertion. The drug is absorbed across the skin into the blood of the circulatory system. Which of the following words most accurately describes the route of administration of the drug?
    a. topical
    b. intravenous
    c. subcutaneous
    d. transdermal

21. Blushing, flushing, and pallor are due to
    a. deposition of melanin in the epidermal cells.
    b. changes in blood flow through the dermal blood vessels.
    c. the rate of keratinization of the epidermal cells.
    d. burning of the skin as in sunburn.

22. Cyanosis is
    a. indicative of hypoxemia.
    b. genetically determined.
    c. caused by contraction of the arrector pili muscles.
    d. prevented by the daily ingestion of iron supplements.

23. Attached to the hair follicle is a group of muscles
    a. called the arrector pili muscles.
    b. that cause "goose bumps" when contracted.
    c. that contract in response to fear and exposure to cold temperatures.
    d. All of the above are true

24. At birth, a newborn is covered by vernix caseosa, which is secreted by
    a. the placenta.
    b. the sebaceous glands of the newborn.
    c. the eccrine glands of the newborn.
    d. the ceruminous glands of the mother.

25. With which skin function are the words *core* and *shell* associated?
    a. thermoregulation
    b. vitamin D synthesis
    c. skin color determination
    d. tactile (touch) sensation

26. Which of the following is implied in the following words: pyrexia, antipyretic, and pyrogenic?
    a. infection
    b. fever
    c. hyperthermia
    d. heat stroke

CASE STUDY

Sixteen-year-old Kevin was involved in a car accident and suffered full-thickness burns over both lower extremities. On admission to the hospital, he was given large amounts of fluid intravenously and placed on sterile sheets.

1. What information can be obtained from the Rule of Nines?
    a. Kevin suffered full-thickness burns.
    b. Kevin was burned over 36% of his body.
    c. Kevin should be experiencing excruciating pain.
    d. Kevin was burned over 75% of his body.

2. What is the primary reason that Kevin requires large amounts of intravenous fluids?
    a. He has not had anything to drink since the accident.
    b. The loss of skin causes the loss of large volumes of body fluid.
    c. Sterile sheets pull water out of the burn site.
    d. He is thirsty.

3. Despite his severe burns, Kevin might not require heavy doses of pain medicine because
    a. he is immobile and not moving the injured site.
    b. the burn site has not yet become infected.
    c. the pain receptors in the dermis have been destroyed.
    d. the melanin within the skin acts as a pain reliever.

Chapter **7** Integumentary System and Body Temperature

PUZZLE

Hint: Red, White, and Blue of Skin

Directions: Perform the following functions on the Sequence of Words that follows. When all the functions have been performed, you are left with a word or words related to the hint. Record your answers in the space provided.

Functions: Remove the following:

1. The two types of sudoriferous glands

2. The two layers of skin

3. The two layers of the epidermis

4. Cells that secrete a dark tanning pigment in response to exposure to UV radiation

5. The layer on which the dermis sits (2 names)

6. The cream cheese–like substance that covers the skin of the newborn infant

7. The pigment that causes jaundice

8. The condition that is caused by a lack of melanin

9. The protein that makes the skin water-resistant

10. Glands most associated with blackheads

11. Words (2) that refer to baldness and excess/abnormal hair growth (often steroid-induced)

12. Words (2) that refer to hives and itching

13. Conditions (2) caused by increases in dermal blood flow.

14. Conditions (2) associated with elevated body temperature

Sequence of Words

URTICARIASUBCUTANEOUSFLUSHINGEPI
DERMISHIRSUTISMSTRATUMCORNEUMERY
THEMAECCRINEPYREXIASEBACEOUSAPO
CRINEDERMISSTRATUMGERMINATIVUMPA
LLORHYPERTHERMIAVERNIXCASEOSAALO
PECIAMELANOCYTESBILIRUBINPRURITISCY
ANOSISHYPODERMISKERATINBLUSHINGA
LBINISM

Answer: _____, _____, _____

8 Skeletal System

Answer Key: Textbook page references are provided as a guide for answering these questions. A complete answer key was provided for your instructor.

OBJECTIVES

1. List the functions of the skeletal system and the classification of bones by size and shape.
2. Differentiate between the composition and location of compact and spongy bone.
3. Describe the structure of a long bone.
4. Describe the roles of osteoblasts and osteoclasts and how bones grow in length and width.
5. List the bones of the axial skeleton, and label important landmarks on selected bones.
6. List the bones of the appendicular skeleton, and label important landmarks on selected bones.
7. List the main types and functions of joints, and describe the types of joint movements.

Part I: Mastering the Basics

MATCHING

Long Bone

Directions: Match the following terms to the most appropriate definition by writing the correct letter in the space provided. See text, pp. 114-117.

A. osteocyte
B. osteoblast
C. red bone marrow
D. compact bone
E. spongy bone
F. haversian system
G. trabeculae
H. diaphysis
I. epiphyseal disc
J. endosteum
K. osteoclast
L. medullary cavity
M. periosteum
N. epiphysis
O. articular cartilage

1. _____ Dense, hard type of bone

2. _____ Bony plates found in spongy bone

3. _____ Band of hyaline cartilage at ends of long bone; longitudinal bone growth occurs here.

4. _____ Tough outer connective tissue covering the diaphysis of a long bone

5. _____ Hollow center of the shaft of a bone

6. _____ Found on the outer surface of the epiphysis at a joint

7. _____ Microscopic unit of compact bone; also called an *osteon*

8. _____ Mature bone cell

9. _____ Inner lining of the medullary cavity of a long bone

10. _____ The shaft of a long bone

11. _____ Site of blood cell formation

12. _____ Eroding activity of this cell remodels and expands the medullary cavity.

13. _____ A bone-building cell

14. _____ Cancellous bone

15. _____ The enlarged end of a long bone

16. _____ This type of bone has a punched-out or "Swiss cheese" appearance.

17. _____ Called the *growth plate*

18. _____ This cell is responsible for bone resorption.

MATCHING

Skull

Directions: Match the following terms to the most appropriate definition by writing the correct letter in the space provided. Some terms may be used more than once. See text, pp. 118-123.

A. maxilla
B. mandible
C. palatine bones
D. frontal
E. parietal
F. external auditory meatus
G. occipital
H. sphenoid
I. zygomatic bones
J. temporal
K. ethmoid
L. fontanels

1. _____ Lower jaw bone; contains the lower teeth

2. _____ Opening in the temporal bone for the ear

3. _____ Upper jaw bone; contains the upper teeth

4. _____ Cheekbones; also form part of the orbits of the eyes

5. _____ Forms the top and sides of the skull

6. _____ Forms the bony structure of the nasal cavity

7. _____ Forms the chin

8. _____ Forms the floor and back wall of the cranium

41

9. _____ The large hole in this bone is called the *foramen magnum.*

10. _____ Forms the posterior part of the hard palate and the floor of the nasal cavity

11. _____ Butterfly-shaped bone that forms part of the floor and sides of the cranium; the sella turcica houses the pituitary gland.

12. _____ Forms the forehead

13. _____ Means "little fountain," the baby's soft spots

14. _____, _____ The coronal suture is the "zipper-appearing" joint between these two bones

15. _____ Contains the external auditory meatus, styloid process, zygomatic process, and mastoid process

16. _____ Articulates with the temporal bone to form a freely movable joint

17. _____ The condyles of this bone sit on the atlas.

18. _____ The M in TMJ

19. _____ The T in TMJ

20. _____ Bone that articulates with the parietal bone at the coronal suture

READ THE DIAGRAM

The Skull

Directions: Referring to the diagram, write the numbers in the blanks below. Some numbers on the skull may not be used. See text, pp. 118-122.

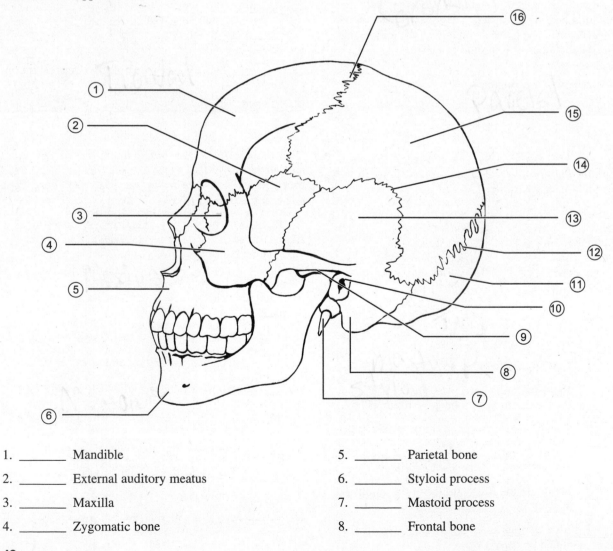

1. _____ Mandible

2. _____ External auditory meatus

3. _____ Maxilla

4. _____ Zygomatic bone

5. _____ Parietal bone

6. _____ Styloid process

7. _____ Mastoid process

8. _____ Frontal bone

9. _____ Bone that contains the foramen magnum

10. _____ Bone that contains the external auditory meatus, styloid process, zygomatic process, and mastoid process

11. _____ Bone that articulates with the temporal bone to form a freely movable joint

12. _____ The immovable joint between the frontal and parietal bones

13. _____, _____, _____ Three sutures

14. _____ Ethmoid bone

15. _____ Sphenoid bone

16. _____ TMJ

MATCHING

Thoracic Cage

Directions: Match the following terms to the most appropriate definition by writing the correct letter in the space provided. Some terms may be used more than once. See text, pp. 126-127.

A. false ribs
B. true ribs
C. floating ribs
D. manubrium
E. xiphoid process
F. body
G. costal angle
H. manubriosternal joint
I. suprasternal notch
J. costal margin
K. left midclavicular line

1. _____ First seven pairs of ribs

2. _____ Next five pairs of ribs

3. _____ Last two pairs of false ribs

4. _____ Lower tip of the sternum

5. _____ The ribs that are closest to the clavicle

6. _____ The depression on the superior border of the manubrium

7. _____ Ribs that attach directly to the sternum by the costal cartilage

8. _____ The part of the sternum closest to the collar bone

9. _____ Ribs that do not attach to the sternum

10. _____ Ribs that attach indirectly to the sternum

11. _____ The largest part of the sternum; located between the manubrium and the xiphoid process

12. _____ The articulation between the manubrium and the body of the breastbone

13. _____ Also called the *angle of Louis*

14. _____ Rib #2 is located at this articulation.

15. _____ Should be less than 90 degrees

16. _____ Imaginary line parallel to the midsternal line and extends downward from the collar bone

MATCHING

Bones Forming the Shoulder Girdle and Upper Extremities

Directions: Match the following terms to the most appropriate definition by writing the correct letter in the space provided. Some terms may be used more than once. See text, pp. 128-130.

A. humerus
B. phalanges
C. glenoid cavity
D. scapula
E. ulna
F. pectoral girdle
G. carpals
H. pollex
I. acromion
J. metacarpals
K. olecranon process
L. radius
M. clavicle

1. _____ Shoulder blade or wing bone

2. _____ Depression where the head of the humerus articulates with the scapula

3. _____ Long bone of the arm that articulates with the scapula

4. _____ Bony point of the ulna that forms the elbow

5. _____ Collar bone

6. _____ Bones that form the palm of the hand

7. _____ The 14 bones that form the fingers

8. _____ Bone that contains the glenoid cavity

9. _____ The clavicle and scapula form this structure.

10. _____ The scapula articulates with this bone to form a ball-and-socket joint at the shoulder.

11. _____ The ulna and this bone articulate to form a hinge joint at the elbow.

12. _____ Bone that contains the olecranon fossa

13. _____ Long bone in the forearm that is located on the same side as the little finger

14. _____ Wrist bones

15. _____ Bone in the forearm that is on the thumb side

16. _____ Process on the scapula that is the pointy part of the shoulder

17. _____ Also called the *shoulder girdle*

18. _____ "Sidekick" bone of the radius

19. _____ Contains the acromion and coracoid process

20. _____ Bones that articulate with the proximal phalanges

21. _____ Bone that contains the olecranon process

22. _____ The radius and this bone "cross" during pronation.

23. _____ Long slender bone that articulates with both the sternum and the scapula

24. _____ The thumb

25. _____ The phalange that contains only two bones

26. _____ Bones that articulate with the metacarpals and distal radius and ulna

27. _____ The scapula and this bone form the shoulder girdle.

READ THE DIAGRAM

Shoulder and Upper Extremities

Directions: Refer to the diagram and write the numbers in the blanks. Some numbers will be used more than once. See text, pp. 128-129.

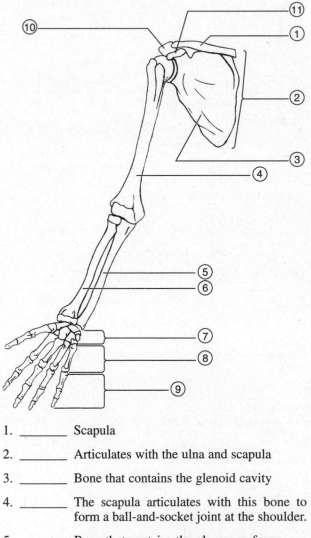

7. _____ Small bones that are distal to the radius and proximal to the metacarpals

8. _____ Radius

9. _____ Pectoral girdle

10. _____ The acromion

11. _____ Coracoid process

12. _____ Bone that contains the olecranon process

13. _____ The radius and this bone "cross" during pronation.

14. _____ Long, slender bone that articulates with both the sternum and the scapula

MATCHING

Bones of the Pelvic Girdle and Lower Extremities

Directions: Match the following terms to the most appropriate definition by writing the correct letter in the space provided. Some terms may be used more than once. See text, pp. 130-133.

A. coxal bone
B. symphysis pubis
C. trochanter
D. tibia
E. metatarsals
F. calcaneus
G. fibula
H. acetabulum
I. obturator

J. ilium
K. ischial tuberosity
L. patella
M. femur
N. pelvic girdle
O. phalanges
P. hallux
Q. talus

1. _____ The part of the coxal bone on which you sit

2. _____ Bone that contains the obturator foramen

3. _____ Kneecap

1. _____ Scapula

2. _____ Articulates with the ulna and scapula

3. _____ Bone that contains the glenoid cavity

4. _____ The scapula articulates with this bone to form a ball-and-socket joint at the shoulder.

5. _____ Bone that contains the olecranon fossa

6. _____ Ulna

44

4. _____ Shin bone

5. _____ Cartilaginous disc between the two pubic bones; the disc expands during pregnancy.

6. _____ The part of the hip bone that "flares"

7. _____ The instep of the foot is formed by these bones.

8. _____ Thin long bone of the leg; bears less body weight than the shin bone

9. _____ Heel bone

10. _____ Large bony process on the femur

11. _____ The "next of shin"

12. _____ Largest bone in the body; called the *thigh bone*

13. _____ The head of the femur articulates with this depression in the coxal bone.

14. _____ The cuplike depression formed by the union of the ilium, ischium, and pubis

15. _____ Formed by the two coxal bones

16. _____ The distal end of this bone is the lateral malleolus.

17. _____ Articulates with the coxal bone to form the hip and the tibia to form the knee

18. _____ The distal end of this bone is the medial malleolus.

19. _____ Composed of the pubis, ischium, and ilium

20. _____ Another name for the great toe

21. _____ This tarsal bone articulates with both the tibia and fibula.

22. _____ The long weight bearing bone in the leg

23. _____ Forms the ball of the foot

24. _____ The two-bone phalange

READ THE DIAGRAM

Hip and Lower Extremity

Directions: Refer to the diagram and then write the numbers in the blanks. See text, pp. 130-133.

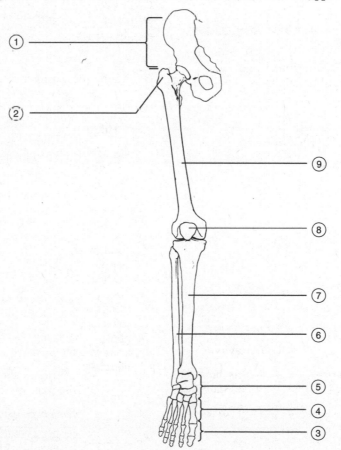

1. _____ Patella

2. _____ Shin bone

3. _____ Instep of the foot is formed by these bones.

4. _____ Thin long bone of the leg; bears less body weight than the shin bone

5. _____ Bony process on the proximal femur

6. _____ Phalanges

7. _____ Largest bone in the body; called the *thigh bone*

8. _____ Contains the acetabulum

9. _____ Distal end of this bone is the lateral malleolus

10. _____ Articulates with the coxal bone to form the hip and the tibia to form the knee

11. _____ Distal end of this bone is the medial malleolus.

12. _____ Composed of the pubis, ischium, and ilium

13. _____ The long weight-bearing bone in the leg

MATCHING

Bone Markings

Directions: Indicate the bone on which each marking occurs. See text, pp. 118-133.

A. femur
B. ulna
C. temporal
D. tibia
E. humerus
F. scapula
G. occipital
H. sternum
I. fibula
J. coxal
K. axis
L. sphenoid

1. _____ Lateral malleolus

2. _____ Foramen magnum

3. _____ Zygomatic process

4. _____ Medial malleolus

5. _____ Greater sciatic notch

6. _____ Obturator foramen

7. _____ Olecranon process

8. _____ Symphysis pubis

9. _____ Greater trochanter

10. _____ Acromion process

11. _____ External auditory meatus

12. _____ Iliac crest

13. _____ Olecranon fossa

14. _____ Glenoid cavity

15. _____ Acetabulum

16. _____ Ischial tuberosity

17. _____ Odontoid process ("dens")

18. _____ Mastoid process

19. _____ Sella turcica

20. _____ Xiphoid process

21. _____ Coracoid process

22. _____ Lesser trochanter

23. _____ Jugular notch

MATCHING

Movement and Joints

Directions: Match the following terms to the most appropriate definition by writing the correct letter in the space provided. Some terms may be used more than once. See text, pp. 133-139.

A. suture
B. ball-and-socket
C. flexion
D. hinge
E. pronation
F. supination
G. extension
H. slightly movable joints
I. dorsiflexion
J. adduction
K. circumduction
L. plantar flexion
M. abduction
N. bursae
O. ligament

1. _____ The type of freely movable joint at the elbow

2. _____ The type of joint formed by the acetabulum and the head of the femur

3. _____ The type of joint that connects the frontal and parietal bones

4. _____ Small sacs of synovial fluid that ease movement at the joint

5. _____ The symphysis pubis and intervertebral discs, for example

6. _____ The type of joint formed by the head of the humerus and the glenoid cavity

7. _____ Turning the forearm so that the palm is facing the sky

8. _____ Describes the tiptoe position, relative to the foot

9. _____ Describes bending the foot toward the shin

10. _____ The type of joint formed by the distal finger bones

11. _____ A circular type of movement made at a ball-and-socket joint such as the shoulder (as in pitching a softball)

12. _____ Movement away from the midline of the body

13. _____ Type of movement achieved as the forearm bends toward the arm (decreasing the angle at the joint)

14. _____ Tough strands of connective tissue that connect bone to bone

15. _____ Straightening of a joint so that the angle between bones increases

16. _____ Movement toward the midline of the body

17. _____ Turning the forearm so that the hand is facing downward (toward the ground)

18. _____ Example of an immovable joint

Draw-A-Line . . . The Joints

Directions: At right is a list of joints. Draw a line to each joint on the skeleton. Use several different colors.

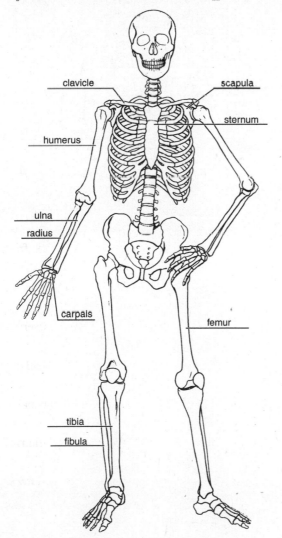

1. Right humeroulnar joint

2. Right proximal tibiofibular joint

3. Left manubriosternal (or sternomanubrial) joint

4. Left temporomandibular joint

5. Left sternoclavicular joint

6. Left acromioclavicular joint

7. Right humeroradial joint

8. Left metacarpophalangeal joint

9. Left tibiofemoral joint

10. Symphysis pubis

11. Left glenohumeral joint

12. Left radiocarpal joint

13. Right distal tibiofibular joint

14. Right proximal tibiofibular joint

15. Left carpometacarpal joint

16. Right proximal radioulnar joint

17. Right distal radioulnar joint

18. Left metatarsophalangeal joint

19. Left talocrural joint

READ THE DIAGRAM

The Skeleton

Directions: Referring to the diagram on p. 48, fill in the blanks with the correct number

1. _____ Bone that articulates with the radius and humerus

2. _____ Bones that are distal to the radius and ulna and proximal to the metacarpals

3. _____ Head of this large bone that articulates with the acetabulum of the coxal bone

4. _____ Bone that is lateral to the tibia

5. _____ Bones that are distal to the tibia and fibula and proximal to the metatarsals

6. _____ Upper extremity phalanges

7. _____ Bone that contains the "pointy" part of the elbow

8. _____ Bone that is composed of the ilium, ischium, and pubis

9. _____ The ulna and this bone supinate and pronate.

10. _____ Bone that is composed of the manubrium, body, and xiphoid process

11. _____ Bone that contains the greater and lesser trochanter and articulates with the coxal bone

12. _____ The scapula and this bone form the pectoral girdle.

13. _____ Bone that contains the glenoid cavity and allows ball-and-socket movement of the arm

14. _____ Bone that is called the *wing bone*

15. _____ Bone that contains the obturator foramen

16. _____ Thigh bone; the largest bone in the body

17. _____ The distal end of this bone is called the *lateral malleolus.*

18. _____ The distal end of this bone is called the *medial malleolus.*

19. _____ Bones that form the palms of the hands

20. _____ The ulna and this bone form a hinge joint at the elbow.

21. _____ Long bone in the forearm that is located on the little finger side of the arm

22. _____ Talus

23. _____ Calcaneus

READ THE DIAGRAM

The Skeleton

Directions: Referring to the diagram below, indicate the name of the bone and whether it is part of the appendicular or axial skeleton. Record your answers on the table on p. 49. Put the answers in the lines opposite the bone number. (Bone 12, the femur, appendicular skeleton, is done for you.)

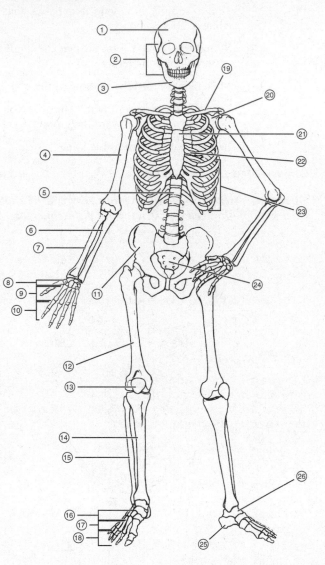

Chapter **8** **Skeletal System**

Bone Number	Bone Name	Skeleton Part
12	Femur	Appendicular
15		
6		
16		
13		
24		
21		
19		
3		
7		
14		
25		
4		
20		
11		
26		

Proximals and Distals

Directions: For each question, determine the most proximal and most distal bone or structure. Place a P over the structure that is most PROXIMAL and a D over the structure that is most DISTAL.

1. Humerus, carpals, olecranon process, radius

2. Femur, tibial tuberosity, talus, distal fibula

3. Lateral malleolus, tibial tuberosity, greater trochanter, patella

4. Hallux, medial malleolus, calcaneus, proximal fibula

5. Phalanges (upper extremities), styloid process (ulna), olecranon fossa, carpals

6. Lateral malleolus, proximal tibiofibular joint, patella, calcaneus

7. Hallux, medial malleolus, talus, femur

8. Styloid process (radius), olecranon process, proximal radioulnar joint, humerus

BODY TOON

Hint: An Indonesian Volcano

Answer: cracka-TOE-a (Krakatoa)

Chapter **8** **Skeletal System**

Locate It: Directional Terms

Directions: Place the letters on the diagram of the skeleton.

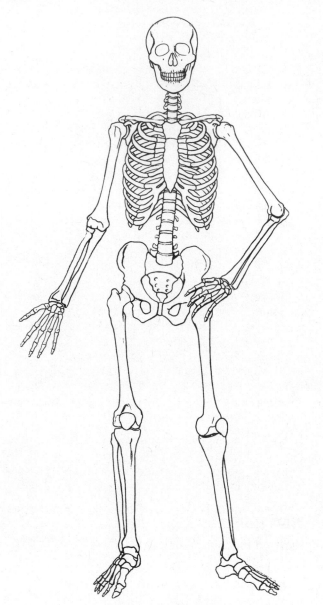

1. Place the letter A on a point distal to the right patella and proximal to the lateral malleolus.

2. Place the letter B on the lateral surface of the left femur and distal to the greater trochanter.

3. Place the letter C on the lateral surface of the femur on a point distal to the right greater trochanter.

4. Place the letter D on a point that is proximal to the right olecranon process and distal to the head of the humerus.

5. Place the letter E on a point distal to the left humerus and proximal to the left carpals.

6. Place the letter F on the right iliac crest.

7. Place the letter G midway between the right glenoid cavity and the olecranon fossa.

8. Place the letter H on a point inferior to the manubrium and superior to the xiphoid process.

9. Place the letter I over the left ischial tuberosity.

10. Place the letter J over an area that is superior to the sacrum and inferior to the T12 vertebra.

11. Place the letter K on a point between the two ischial tuberosities.

12. Place the letter L on the point that is distal to the right lateral malleolus and proximal to the hallux.

13. Place the letter M on the medial right thigh.

14. Place the letter N on the distal right radius.

15. Place the letter O on the bone that is medial to the right radius.

16. Place the letter P on a point immediately distal to the styloid process of the right ulna and proximal to the phalanges.

17. Place the letter Q on the medial side of the left thigh bone.

18. Place the letter R on a point that is superior to the atlas and axis.

19. Place the letter S over a point that is superior to the zygomatic bone.

20. Place the letter T to indicate a point immediately distal to the left humeroulnar joint.

21. Place the letter U over the right acromioclavicular joint.

22. Place the letter V over the left humeroulnar joint.

23. Place the letter W over the symphysis pubis.

24. Place the letter X over the manubriosternal joint.

25. Place the letter Y on the facial bone that articulates with the temporal bone.

26. Place the letter Z over the suprasternal notch.

Bones to Color and Questions to Answer

Directions: Using the diagram of the skeleton, color all structures mentioned in the stem of the questions below. Use different colors for the bones identified in each question. If the structures are not visible on the skeleton, merely draw an arrow pointing to the structure (e.g., temporomandibular joint). Answer the multiple-choice questions AFTER you do the coloring exercise. The answers can be determined by observing the colored bones or bony markings.

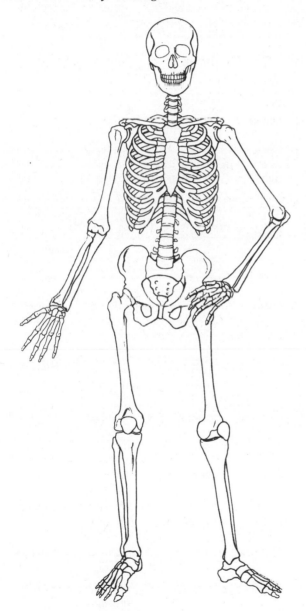

1. What do these three bones have in common—the humerus, ulna, and radius?
   a. All articulate at the wrist.
   b. All articulate at the elbow.
   c. All articulate at the glenoid cavity.
   d. All form a ball-and-socket joint.

2. What do these three structures have in common—acromion, coracoid process, and greater trochanter? All are
   a. located on the scapula.
   b. found on bones of the appendicular skeleton.
   c. located on bones of the axial skeleton.
   d. irregular bones.

3. What do the following three structures have in common—glenoid cavity, olecranon fossa, and acetabulum?
   a. All are bony depressions that help form joints.
   b. All are bony processes to which muscles attach.
   c. All are located on bones that form the axial skeleton.
   d. All are located on long bones.

4. What do the following three structures have in common—pubis, ilium, and ischium?
   a. All are located in the upper extremities.
   b. All are long bones.
   c. All are parts of the coxal bone.
   d. All are bony depressions.

5. What do the following three structures have in common—elbow, knee, and interphalangeal joints?
   a. All are formed exclusively by short bones.
   b. All are ball-and-socket joints.
   c. All perform circumduction movement.
   d. All are hinge joints.

6. What do the following three bones have in common—femur, humerus, and phalanges?
   a. All are located in the lower extremity.
   b. All articulate with the acetabulum.
   c. All form ball-and-socket joints.
   d. All are long bones.

7. What do the following three bones have in common—occipital, zygomatic, and frontal?
   a. There are two of each (they are paired bones).
   b. They are cranial bones.
   c. They are long bones.
   d. They are bones of the skull.

8. What do the following three bones have in common—mandible, maxilla, and zygomatic?
   a. All articulate at the TMJ.
   b. All "hold" teeth.
   c. All are cranial bones.
   d. All are facial bones.

9. What do the following three bones have in common: carpals, ulna, and radius?
   a. All are long bones.
   b. All are proximal to the humerus.
   c. All are bones of the appendicular skeleton.
   d. All articulate at the elbow.

10. What do the following three structures have in common—fibula, femur, and tibia?
    a. All are distal to the patella.
    b. All articulate with the acetabulum.
    c. All are bones of the axial skeleton.
    d. All are proximal to the tarsals.

11. What do the following three structures have in common—sternum, true ribs, and false ribs?
    a. All are long bones.
    b. All are located inferior to the diaphragm.
    c. All help form the thoracic cage.
    d. All articulate with the lumbar vertebrae.

12. What do the following three structures have in common—manubrium, body, and xiphoid process?
    a. All articulate with all 12 pairs of ribs.
    b. All articulate with the thoracic vertebrae.
    c. All articulate with the clavicle.
    d. All are parts of the sternum.

13. What do the following three structures have in common—lateral malleolus, medial malleolus, and tibial tuberosity?
    a. All are bony processes on the tibia.
    b. All are bony processes on the fibula.
    c. All articulate with the acetabulum.
    d. All are bony processes found on the bones of the appendicular skeleton.

14. Identify the following bone. It articulates distally with the carpal bones and proximally with the humerus.
    a. phalange
    b. scapula
    c. clavicle
    d. ulna

15. Identify the following bone or bony structure. It articulates distally with the tibia and proximally with the coxal bone.
    a. fibula
    b. acetabulum
    c. ischium
    d. femur

16. Identify the following bone. It articulates distally with the tarsal bones and proximally with the femur.
    a. tibia
    b. calcaneus
    c. coxal bone
    d. metatarsal

SIMILARS AND DISSIMILARS

Directions: Circle the word in each group that is least similar to the others. Indicate the similarity between the three words on the line below each question.

1. osteoblast   osteoma   osteocyte   osteoclast

   _____

2. compact   spongy   suture   cancellous

   _____

3. maxilla   patella   mandible   palatine bones

   _____

4. frontal   clavicle   parietal   occipital

   _____

5. atlas   vertebra   wrist   axis

   _____

6. suture   diaphysis   epiphysis   medullary cavity

   _____

7. fossa   foramen   articulation   meatus

   _____

8. trochanter   epicondyle   pivot   tuberosity

   _____

9. sphenoid   temporal   parietal   scapula

   _____

10. true   floating   manubrium   false

   _____

11. manubrium   ribs   body   xiphoid process

   _____

12. scapula   pectoral girdle   olecranon process   clavicle

   _____

13. pubis   xiphoid process   ilium   ischium

_____

14. femur   glenoid cavity   tibia   fibula

_____

15. tarsals   phalanges   carpals   metatarsals

_____

16. radius   metatarsals   ulna   humerus

_____

17. zygomatic   mastoid   foramen   auditory
    process   process   magnum   meatus

_____

18. foramen   coronal   lambdoidal   sagittal

_____

19. glenoid   olecranon   greater   acetabulum
    cavity   fossa   trochanter

_____

20. acromion   glenoid   coracoid   olecranon
                cavity   process   fossa

_____

21. hallux   symphysis pubis   talus   calcaneus

_____

22. hinge   fontanel   pivot   ball-and-socket

_____

23. extension   suture   abduction   plantar
                                         flexion

_____

24. synovial fluid   articulation   hinge   joint

_____

25. long   meatus   short   irregular

_____

26. kyphosis   osteoporosis   scoliosis   lordosis

_____

27. cervical   thoracic   lumbar   paranasal

_____

28. periosteum   hallux   endosteum   trabeculae

_____

29. haversian   osteon   articulation   osteocyte

_____

30. pollex   elbow   phalanges   thumb

_____

31. spine   lamina   vertebral foramen   foramen
                                            magnum

_____

32. axis   dens   C2   sacrum

_____

33. clavicle   tailbone   coccyx   sacrum

_____

34. supination   "palms   fontanel   pronation
                  down"

_____

READ THE DIAGRAM

Movements

Directions: Referring to the illustration on the next page, fill in the blanks with the words listed below.

A. flexion
B. extension
C. inversion
D. adduction
E. supination
F. eversion
G. circumduction
H. hyperextension
I. pronation
J. plantar flexion
K. dorsiflexion
L. abduction

1. _____ Movement from 1 → 2

2. _____ Movement from 3 → 4

3. _____ Movement from 5 → 6

4. _____ Movement from 8 → 7

5. _____ Movement from 9 → 10

6. _____ Movement from 13 → 12

7. _____ Movement from 12 → 11

8. _____ Movement from 2 → 1

9. _____ Movement from 4 → 3

10. _____ Movement from 6 → 5

11. _____ Movement from 7 → 8

12. _____ Movement from 10 → 9

13. _____ Movement from 11 → 13

14. _____ Movement from 12 → 13

15. _____ Movement illustrated by 14

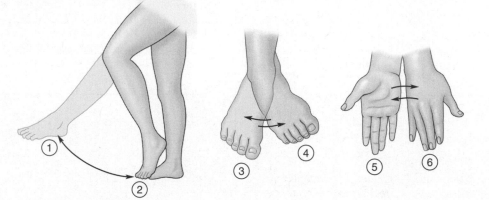

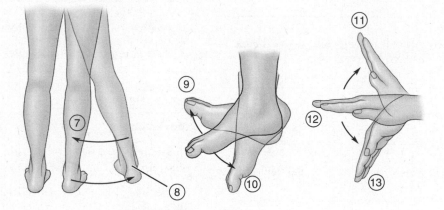

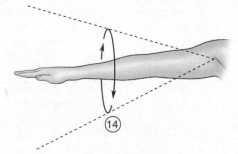

READ THE DIAGRAM

Movements

Directions: Referring to the illustration on the previous page, fill in the number described below.

1. _____ Illustrates the pronated extremity

2. _____ Illustrates the extremity in plantar flexion

3. _____ Illustrates the everted extremity

4. _____ Illustrates the movement of a ball-and-socket joint

5. _____ Illustrates the effects of the crossing of the ulna and radius

6. _____ Illustrates dorsiflexion of the extremity

7. _____ Illustrates the movement of the lower extremity toward the midline on the body

8. _____ Illustrates movement of the head of the humerus within the glenoid cavity

9. _____ Illustrates movement of the head of the femur within the acetabulum; thigh abducts

BODY TOON

Hint: Name This Bone in the Appendicular Zone

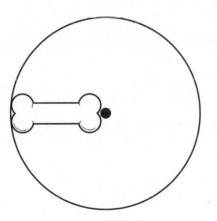

Part II: Putting It All Together

MULTIPLE CHOICE

Directions: Choose the correct answer.

1. The purpose of the articular cartilage is to
   a. provide oxygenated blood to the osteon (haversian system).
   b. reduce friction within the joint.
   c. produce blood cells.
   d. act as a site of attachment for ligaments.

2. Injury to the epiphyseal disc is most likely to
   a. impair longitudinal bone growth.
   b. suppress blood cell formation, causing anemia.
   c. cause bone softening and bowing of the legs.
   d. deprive the bone of oxygenated blood.

3. Which of the following is true of the periosteum?
   a. lines the medullary cavities
   b. makes blood cells
   c. covers the outside of the diaphysis
   d. covers the distal end of long bones to minimize friction within the joints

4. Parathyroid hormone
   a. lowers blood levels of calcium.
   b. stimulates the bone marrow to make blood cells.
   c. stimulates bone formation.
   d. stimulates osteoclastic activity.

5. Which of the following is most likely to increase bone density?
   a. excess parathyroid hormone activity
   b. osteoblastic activity
   c. bone marrow depression
   d. prolonged bed rest

6. Which of the following is least true of the shoulder girdle?
   a. is also called the *pectoral girdle*
   b. includes the scapulae and clavicles
   c. supports the weight of the skull
   d. attaches the upper extremities to the axial skeleton

7. Which of the following is true of the femur?
   a. contains the olecranon process
   b. articulates with the coxal bone at the acetabulum
   c. articulates with both the coxal bone and the coccyx
   d. is also called the *shin bone*

Answer: radius

Chapter **8** **Skeletal System**

8. The lateral malleolus and medial malleolus are located on the
   a. distal tibia.
   b. distal fibula.
   c. proximal and distal tibia.
   d. distal leg bones.

9. Which of the following is not characteristic of vertebrae?
   a. stacked in a column and separated by discs of cartilage
   b. include the atlas and axis
   c. include the sacrum and coccyx
   d. include the body, manubrium, and xiphoid process

10. Fontanels and unfused sutures in the infant skull
    a. are abnormal.
    b. allow for the expansion of the skull for brain growth.
    c. occur only in the presence of maternal deprivation of calcium.
    d. close and fuse within the first month.

11. The cranium
    a. contains all the bones of the skull.
    b. includes the sphenoid bone, the occipital bone, and the zygomatic bone.
    c. houses the brain.
    d. includes the occipital bone, the atlas, and the axis.

12. The occipital bone
    a. contains the obturator foramen.
    b. contains the foramen magnum.
    c. articulates with the cervical prominens.
    d. articulates with the parietal bones at the coronal suture.

13. The foramen magnum
    a. houses the pituitary gland.
    b. allows for the descent of the brainstem as the spinal cord.
    c. is the site of attachment for the large thigh muscles.
    d. forms the eye sockets.

14. This patient has his mandible wired to his maxilla.
    a. an older adult patient who fell and broke his hip
    b. an infant with a developing hydrocephalus (expanding cranial circumference)
    c. an intoxicated young man with a fractured jaw
    d. a middle-aged man with an abscessed tooth

15. The temporal bone
    a. is a facial bone.
    b. contains the mastoid, styloid, and zygomatic processes.
    c. articulates with the maxilla at the TMJ.
    d. contains the foramen magnum.

16. The ulna
    a. articulates with the scapula at the glenoid cavity.
    b. forms a ball-and-socket joint with the radius.
    c. forms a hinge joint with the humerus.
    d. articulates with the tarsals to form the wrist.

17. Which of the following supinate and pronate?
    a. scapula and clavicle
    b. manubrium and xiphoid process
    c. femur and tibia
    d. ulna and radius

18. Which of the following are aligned parallel to each other?
    a. femur and tibia
    b. tibia and calcaneus
    c. fibula and patella
    d. tibia and fibula

19. The acetabulum
    a. is located on the femur.
    b. is formed by the ilium, ischium, and pubis.
    c. is part of the axial skeleton.
    d. forms the knee.

20. Which of the following is not part of the knee?
    a. distal femur
    b. patella
    c. proximal tibia
    d. acetabulum

21. What do the following three structures have in common: diaphysis, epiphysis, and periosteum? They are
    a. types of joints.
    b. parts of an irregular bone.
    c. parts of a long bone.
    d. types of bone marrow.

22. Mr. Phil Anges was named after
    a. his fingers and toes.
    b. a mountain in South America.
    c. a Gucci glove.
    d. his thermal socks.

23. Which of the following is a bone marking?
   a. skin, skin, skin, skin
   b. sight, sight, sight, sight
   c. amen, amen, amen, amen
   d. bode, bode, bode, bode

CASE STUDY

T.B., a 65-year-old woman, fell as she was leaving her home. She was unable to move her left leg and appeared to be in extreme pain when the paramedics transported her to a stretcher. She was later diagnosed with a fractured hip and sent to surgery, where a pin was inserted into the fractured bone. A bone scan indicated a decrease in bone density.

1. Which anatomical structure was fractured?
   a. iliac crest of the coxal bone
   b. ischial tuberosity of the hip bone
   c. greater trochanter of the femur
   d. neck of the femur

2. Given her age and medical history, what does the decrease in her bone density suggest?
   a. cancer of the bone
   b. osteoporosis
   c. congenital defect of the bone
   d. infection of the bone

BODY TOON

Hint: After a Theatrical Performance

Answer: cast party

PUZZLE

Hint: Personal Brain Bucket

Directions: Perform the following functions on the Sequence of Words below. When all functions have been performed, you are left with a word or words related to the hint. Record your answer below.

Functions: Remove the following:

1. Vertebral bones (3)

2. Bones that form the acetabulum (3)

3. Bones that form the pectoral girdle (2)

4. Bones of the leg and ankle (3)

5. Bones that articulate with the phalanges and tarsals

6. Facial bones (4)

7. Bones that form the wrist and palms of the hand

8. Another name for the kneecap

9. Name for the toes and fingers

10. Bones in the forearm that pronate and supinate

11. Arm bone that articulates with the scapula

12. Cranial bones joined by the coronal suture

13. Joint formed by the proximal ulna and the distal humerus

14. Joint formed by the proximal tibia and the distal femur

15. Tarsal bone called the heel bone

16. Joint formed by the mandible and the temporal bone

Chapter 8 **Skeletal System**

Sequence of Words

ILIUMRIBSHUMERUSETHMOIDPARIETALFIBU
LAMANDIBLETMJCARPALSATLASMAXILLAISC
HIUMCLAVICLEELBOWMETACARPALSSACRUM
CALCANEUSULNAZYGOMATICMETATARSALSP
UBISTIBIASCAPULAKNEEPHALANGESFEMURR
ADIUSTARSALSAXISPATELLACRANIUMFRONTAL

Answer: _____

Hint: Rover's Interpretation of the Megalithic Ruin in Wiltshire (England)

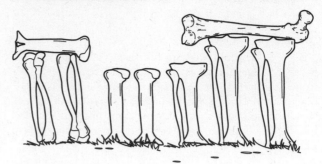

Answer: bone-henge

OBJECTIVES

1. Identify three types of muscle tissues.
2. Compare the structure of the whole muscle and the structure of a single muscle fiber.
3. Describe the sliding filament mechanism of muscle contraction.
4. Explain the role of calcium and adenosine triphosphate (ATP) in muscle contraction.
5. Describe the relationship between skeletal muscles and nerves, including the motor unit and its relationship to recruitment and the events that occur at the neuromuscular junction.
6. Discuss the force of muscle contraction, including the following:
   - Define *twitch* and *tetanus*.
   - Identify the sources of energy for muscle contraction.
   - Trace the sequence of events from nerve stimulation to muscle contraction.
7. Define muscle terms, and state the basis for naming muscles.
8. Identify and list the actions of the major muscles.

Part I: Mastering the Basics

MATCHING

Types of Muscles

Directions: Match the following terms to the most appropriate definition by writing the correct letter in the space provided. Some terms may be used more than once. See text, pp. 143, 144.

A. skeletal muscle
B. smooth muscle
C. cardiac muscle

1. _____ Striated and voluntary

2. _____ Found in the walls of organs or viscera

3. _____ Found in the walls of blood vessels

4. _____ Striated and involuntary

5. _____ Found in the heart

6. _____ Nonstriated and involuntary

7. _____ Found attached to bones

8. _____ The only type of muscle that is nonstriated

9. _____ The only type of muscle that is voluntary

10. _____ Must be innervated by a somatic motor nerve

MATCHING

Parts of a Whole Muscle

Directions: Match the following terms to the most appropriate definition by writing the correct letter in the space provided. Some terms may be used more than once. See text, pp. 144-146.

A. fascicles
B. perimysium
C. endomysium
D. aponeurosis
E. tendon
F. epimysium

1. _____ Cordlike structure that attaches muscle to bone

2. _____ Small bundles of muscle fibers

3. _____ Connective tissue that surrounds individual muscle fibers

4. _____ Flat sheetlike fascia that attaches muscle to muscle or muscle to bone

5. _____ Connective tissue that surrounds the fascicles or bundles

6. _____ Connective tissue that surrounds a whole skeletal muscle

MATCHING

Structure of a Muscle Fiber

Directions: Match the following terms to the most appropriate definition by writing the correct letter in the space provided. See text, pp. 145-147.

A. T tubule
B. sarcoplasmic reticulum (SR)
C. actin
D. sarcomere
E. myosin
F. cross-bridge

1. _____ Series of contractile units that make up each myofibril; extends from Z line to Z line

2. _____ Thin protein filaments that extend toward the center of the sarcomere from the Z lines

3. _____ Thick protein filaments whose "heads" form cross-bridges when they interact with the thin filaments

4. _____ Calcium is stored within this structure in the relaxed muscle.

5. _____ An extension of the sarcolemma or cell membrane that penetrates the interior of the muscle; the electrical signal runs along this membrane toward the sarcoplasmic reticulum.

6. _____ The temporary connection formed when the myosin heads interact with the actin, causing muscle contraction

7. _____ The troponin-tropomyosin complex and this protein are called the thin filament.

ORDERING

Muscle Stimulation, Contraction, and Relaxation

Directions: Using numbers 1 to 8, place the following events in the correct order. The first and fourth events are labeled for you. See text, pp. 146-151.

1. _____ The electrical signal (nerve impulse) causes the vesicles within the nerve terminal to fuse with the membrane and release the neurotransmitter acetylcholine (ACh).

2. _____ Muscle relaxation occurs when calcium is pumped back into the sarcoplasmic reticulum.

3. _____ The electrical signal in the muscle membrane travels along the sarcolemma (muscle membrane) and penetrates deep into the muscle by the T tubular system.

4. _____ ACh stimulates the receptors and causes an electrical signal to develop along the muscle membrane.

5. _____ Stimulation of the motor nerve causes an electrical signal to move along the somatic nerve toward the nerve ending.

6. _____ The calcium allows the actin, myosin, and ATP to form cross-bridges, thereby causing muscle contraction.

7. _____ ACh diffuses across the junction and binds to the receptor sites on the muscle membrane.

8. _____ The electrical signal stimulates the SR to release calcium into the sarcomere area.

MATCHING

Muscle Terms

Directions: Match the following terms to the most appropriate definition by writing the correct letter in the space provided. Some terms may be used more than once. See text, pp. 153, 154.

A. origin
B. insertion
C. antagonist
D. hypertrophy
E. synergist
F. prime mover
G. atrophy
H. contracture

1. _____ The muscle responsible for most of the movement in a group of muscles; called the *chief muscle*

2. _____ Muscle attachment to the movable bone

3. _____ An increase in the size of a muscle because of overuse

4. _____ Muscle attachment to the stationary bone

5. _____ Helper muscle; works with other muscles to produce the same movement

6. _____ Abnormal formation of fibrous tissue in muscles, preventing normal mobility

7. _____ Muscle that opposes the action of another muscle

8. _____ Wasting away or decrease in the size of the muscles

9. _____ Classified as disuse, denervation, and senile

MATCHING

Naming Skeletal Muscles

Directions: Match the following terms to the most appropriate definition by writing the correct letter in the space provided. Some terms may be used more than once. See text, p. 154.

A. vastus
B. deltoid
C. brevis
D. rectus
E. pectoralis
F. brachii
G. gluteus
H. maximus
I. teres
J. latissimus
K. oblique

1. _____ Short

2. _____ Diagonal

3. _____ Triangular

4. _____ Round

5. _____ Buttock

6. _____ Large

7. _____ Huge

8. _____ Arm

9. _____ Wide

10. _____ Straight

11. _____ Chest

MATCHING

Muscles of the Head and Neck

Directions: Match the following terms to the most appropriate definition by writing the correct letter in the space provided. Some terms may be used more than once. See text, pp. 154-161.

A. frontalis
B. orbicularis oculi
C. orbicularis oris
D. buccinator
E. masseter
F. temporalis
G. sternocleidomastoid
H. zygomaticus
I. trapezius
J. platysma
K. levator palpebrae superioris

1. _____ Muscle that flattens the cheek when contracted; positions food for chewing

2. _____ Flat muscle that raises the eyebrows and wrinkles the forehead; creates a surprised look

3. _____ Fan-shaped muscle that extends from the temporal bone to the mandible; works synergistically with other chewing muscles

4. _____ Extends from the corners of the mouth to the cheekbone; called the *smiling muscle*

5. _____ Muscle of the upper back and neck. Contraction of this muscle tilts the head so that the face looks up at the sky; also shrugs the shoulder

6. _____ Sphincter muscle encircling the mouth; called the *kissing muscle*

7. _____ Muscles on either side of the neck that cause flexion of the head; contraction of one of these muscles rotates the head to the opposite side.

8. _____ A chewing muscle that works synergistically with the temporalis muscle

9. _____ Sphincter muscle encircling the eyes; assists in winking, blinking, and squinting

10. _____ Works antagonistically to the sternocleidomastoid

11. _____ A spasm of this muscle causes torticollis, or wryneck.

12. _____ Attaches to the collar bone, breast bone, and temporal bone

13. _____ Muscle that raises the eyelid

14. _____ Your frowning and pouting muscle; also depresses the mandible

MATCHING

Muscles of the Trunk and Extremities

Directions: Match the following terms to the most appropriate definition by writing the correct letter in the space provided. Some terms may be used more than once. See text, pp. 160-166.

A. intercostals
B. abdominal muscles
C. linea alba
D. adductors
E. triceps brachii
F. sartorius
G. trapezius
H. serratus anterior
I. hamstrings
J. pectoralis major
K. biceps brachii
L. quadriceps femoris
M. latissimus dorsi
N. gastrocnemius
O. gluteus maximus
P. deltoid
Q. diaphragm
R. tibialis anterior
S. soleus
T. rotator cuff muscles

1. _____ Dome-shaped muscle that separates the thoracic cavity from the abdominal cavity

2. _____ The chief muscle of inhalation (breathing in)

3. _____ Longest muscle in the body; used to sit cross-legged

4. _____ Muscle that shrugs the shoulders and hyperextends the head

5. _____ Barbecued ribs

6. _____ Jagged muscle that resembles the teeth of a saw; lowers the shoulder and moves the arm, as in pushing a cart

7. _____ Includes the internal oblique, external oblique, transversus, and rectus

8. _____ Muscles located on the medial (inner) surface of the thigh; horseback riders use these muscles to grip the horse with their thighs.

9. _____ Group of muscles that extend or straighten the leg at the knee, as in kicking a football

10. _____ Large, broad muscle located over the middle and lower back; lowers the shoulders and brings the arms back, as in swimming and rowing

11. _____ The Achilles tendon attaches the soleus and this muscle to the heel bone.

12. _____ Large, broad muscle that forms the anterior chest wall; connects the humerus with the clavicle and structures of the chest

13. _____ Muscle that forms the shoulder pads; positions the arms in a "scarecrow" position

14. _____ Muscles responsible for raising the rib cage during inhalation (breathing)

15. _____ A white line that extends from the sternum to the pubic bone; formed by the aponeurosis of the abdominal muscles

16. _____ Muscle that flexes the forearm; when you ask a child to "make a muscle," this is the one that pops up.

17. _____ Muscle that lies along the posterior surface of the humerus; it extends the forearm and is used to bear weight in crutch-walking.

18. _____ Largest muscle in the body that forms part of the buttocks; you sit on this muscle.

19. _____ The calf muscle that is used in plantar flexion; also called the *toe dancer's muscle*

20. _____ Muscle group on the posterior surface of the thigh that extends the thigh at the hip and flexes the leg at the knee; antagonist to the quadriceps femoris

21. _____ With regard to plantar flexion, works synergistically with the gastrocnemius

22. _____ Includes the rectus femoris, vastus medialis, vastus lateralis, and vastus intermedius

23. _____ Located over the shin bone; causes dorsiflexion

24. _____ Includes the biceps femoris

25. _____ The brachialis and the brachioradialis work synergistically with this muscle to flex the forearm at the elbow.

26. _____ Group of four muscles that attach the humerus to the scapula; tendons form a cuff over the proximal humerus

MATCHING

Where Is It Happening: Neuromuscular Junction or Sarcomere?

Directions: For each statement, indicate if it occurs within the neuromuscular junction (NMJ) or the sarcomere (S).

1. _____ Actin, troponin-tropomyosin complex, and myosin

2. _____ N_M blockade

3. _____ Binding of Ca^{2+} by troponin

4. _____ Release of ACh by the vesicles within the axon terminals

5. _____ Sliding of the thick and thin filaments

6. _____ Release of Ca^{2+} from the SR

7. _____ Binding of the ACh to the nicotinic receptors

8. _____ Cross-bridge formation

9. _____ Contraction/relaxation

10. _____ Movement of the tropomyosin away from the myosin binding sites on actin

11. _____ Inactivation of ACh by acetylcholinesterase

12. _____ Pumping of Ca^{2+} into the SR during relaxation

13. _____ Site of action of a drug classified as an anticholinesterase agent

14. _____ Site of action for a curare-like drug

15. _____ From Z line to Z line

16. _____ Place where rigor mortis occurs

READ THE DIAGRAM

Directions: Referring to the diagram (see Fig. 9.7 on pp. 155-156 of the text), indicate if the muscle is best observed on the anterior (A) or posterior (P) view. Answer (A) for anterior view or (P) for posterior view.

1. _____ Pectineus

2. _____ Latissimus dorsi

3. _____ Gastrocnemius

4. _____ Gluteus maximus

5. _____ Sartorius

6. _____ Trapezius

7. _____ Sternocleidomastoid

8. _____ Hamstrings

9. _____ Vastus lateralis, vastus medialis, and rectus femoris

10. _____ Biceps femoris

11. _____ Achilles tendon

12. _____ Linea alba

13. _____ Biceps brachii

14. _____ Rectus abdominis

15. _____ Buccinator

16. _____ Infraspinatus

17. _____ Iliopsoas

18. _____ Pectoralis major

19. _____ Quadriceps femoris

20. _____ Masseter

21. _____ Zygomaticus

22. _____ Calcaneal tendon

23. _____ Triceps brachii

24. _____ Brachioradialis

25. _____ Adductor longus

26. _____ Quadriceps tendon

27. _____ Platysma

SIMILARS AND DISSIMILARS

Directions: Circle the word in each group that is least similar to the others. Indicate the similarity of the three words on the line below each question.

1. troponin-tropomyosin   ACh   actin   myosin

_____

2. smooth   cardiac   fascicle   skeletal

_____

3. hamstrings   Achilles   rectus   quadriceps
                 tendon   femoris   femoris

_____

4. levator   adductor   flexor   rectus

_____

5. semi-   semi-   trapezius   biceps
   tendinosus   membranosus   femoris

_____

6. vastus   longus   tonus   brevis

_____

7. circularis   rectus   extensor   oblique

_____

8. buccinator   zygomaticus   frontalis   latissimus
                                             dorsi

_____

9. rectus   serratus   vastus   vastus
   femoris   anterior   lateralis   medialis

_____

10. quadriceps   trapezius   tibialis   peroneus
    femoris   anterior   longus

_____

11. internal   linea alba   transversus   rectus
    oblique   abdominis   abdominis

_____

12. trapezius   latissimus dorsi   frontalis   gluteus
                                                  maximus

_____

13. hamstrings   latissimus   quadriceps   sartorius
                  dorsi   femoris

_____

14. deltoid   gastrocnemius   soleus   tibialis
                                          anterior

_____

15. pectoralis   serratus   rectus   gastrocnemius
    major   anterior   abdominis

_____

16. masseter   temporalis   gracilis   buccinator

_____

17. soleus   biceps   triceps   brachioradialis
              brachii   brachii

_____

18. fascia   epimysium   sarcoplasmic   perimysium
                           reticulum

_____

19. myofibril   sarcomeres   actin/myosin   tendon

_____

20. NMJ   ACh   acetylcholinesterase   sarcomere

_____

MULTIPLE CHOICE

Directions: Choose the correct answer.

1. Which muscles produce movement of the extremities, maintain body posture, and stabilize joints?
   a. smooth muscles
   b. muscles that are nonstriated and involuntary
   c. skeletal muscles
   d. striated muscles that are involuntary

2. Which muscle is found in the bronchioles (breathing passages) and the blood vessels?
   a. skeletal muscle
   b. muscle that is striated and voluntary
   c. smooth muscle
   d. muscle that is striated and involuntary

3. Which structures slide in the sliding filament mechanism of muscle contraction?
   a. T tubules and the SR
   b. actin and myosin
   c. calcium and ATP
   d. epimysium and perimysium

4. What is the consequence of recruitment?
   a. paralysis
   b. increased force of muscle contraction
   c. depletion of ACh in the neuromuscular junction
   d. inability of the muscle to contract

5. What is the role of the SR in muscle contraction?
   a. synthesizes actin
   b. synthesizes myosin
   c. releases calcium into the sarcomere
   d. releases ACh into the neuromuscular junction

6. What is the consequence of tetanus?
   a. muscle twitching
   b. flaccid paralysis
   c. sustained muscle contraction
   d. atrophy

7. What happens when receptor sites on the muscle membrane are damaged so that the transmitter ACh cannot bind properly?
   a. The muscle hypertrophies.
   b. Muscle contraction is impaired and the patient experiences muscle weakness.
   c. The muscle tetanizes.
   d. The skeletal muscle loses its striations and becomes a smooth muscle.

8. What event causes the muscle to relax?
   a. Calcium is pumped back into the SR.
   b. ATP is pumped into the SR.
   c. The muscle becomes depleted of actin.
   d. The muscle becomes depleted of myosin.

9. What happens when you repetitively and rapidly stimulate a skeletal muscle?
   a. The muscle delivers a single twitch.
   b. A flaccid paralysis develops.
   c. The muscle tetanizes.
   d. The muscle "freezes" and develops a contracture.

10. Which of the following terminates events in the neuromuscular junction?
    a. Calcium is pumped back into the nerve terminal.
    b. The receptors are blocked with actin.
    c. The receptors are blocked with myosin cross-bridges.
    d. ACh is inactivated.

11. What action is antagonistic to the biceps brachii muscle?
    a. The brachialis contracts.
    b. The deltoid contracts.
    c. The triceps brachii contracts.
    d. The brachioradialis contracts.

12. Which condition is most likely to develop in a person whose leg is in a non–weight-bearing cast for several months?
    a. disuse muscle atrophy of the affected leg
    b. hypertrophy of the affected leg
    c. compression of the growth plate, causing the affected leg to be shorter than normal
    d. osteoporosis and pathological fractures

13. Which of the following is indicated by the terms *pectoralis, gluteus, brachii,* and *lateralis*?
    a. location of a muscle
    b. shape of a muscle
    c. size of a muscle
    d. muscle action

14. Which of the following is indicated by the terms *biceps, triceps,* and *quadriceps*?
    a. numbers of origins of the muscle
    b. direction of fibers of the muscle
    c. shape of the muscle
    d. muscle action

15. What is the basis for naming the sternocleidomastoid muscle?
    a. sites of attachment of the muscle
    b. shape of the muscle
    c. size of the muscle
    d. muscle action

16. The hamstrings
    a. extend the leg at the knee (as in kicking a football).
    b. are located on the posterior thigh.
    c. work synergistically with the quadriceps group to adduct the leg.
    d. flex the thigh at the hip.

17. Rigor mortis develops
    a. only in smooth muscle.
    b. in response to a depletion of calcium in the sarcomere.
    c. in response to a buildup of lactic acid and creatine phosphate.
    d. in response to a deficiency of ATP.

18. Creatine phosphate
    a. is an energy source for muscle contraction.
    b. is a contractile protein.
    c. forms the cross-bridges between actin and myosin.
    d. inactivates ACh within the neuromuscular junction.

19. Which of the following is least descriptive of the deltoid muscle?
    a. forms shoulder pads
    b. is a common site of injection of drugs
    c. elevates the arm to a scarecrow position
    d. has its origin on the humerus

20. Why may the weight of bed linen on a bedridden patient cause a disabling contracture?
    a. The linen causes a prolonged dorsiflexion.
    b. The linen adducts the leg.
    c. The linen forces the foot into a position of plantar flexion.
    d. The linen compresses the nerve that supplies the muscles of the feet.

21. Which condition is most associated with aging?
    a. conversion of skeletal muscle to smooth muscle
    b. gradual loss of muscle strength
    c. an overproduction of myoglobin
    d. conversion of smooth muscle to aponeurosis

22. Which of the following is most related to the role of myoglobin?
    a. muscle contraction
    b. synthesis of actin and myosin
    c. storage of calcium
    d. carrier of oxygen

23. Supinators and pronators
    a. are types of smooth muscles.
    b. have their origin on the scapula.
    c. twist the bones of the forearm.
    d. flex the fingers.

24. Which of the following muscles work synergistically?
    a. quadriceps femoris and biceps brachii
    b. brachialis and brachioradialis
    c. gastrocnemius and rotator cuff muscles
    d. pectoralis major and gluteus maximus

25. A motor unit
    a. is located within the sarcomeres.
    b. refers to a motor neuron and the muscle fibers that the neuron supplies.
    c. is connective tissue that wraps muscle fibers into bundles called fascicles.
    d. is found only in visceral muscle.

26. Which group is incorrect?
    a. types of muscles: skeletal, smooth, cardiac
    b. types of striated muscles: skeletal, smooth, cardiac
    c. muscles involved in breathing: diaphragm, intercostals
    d. muscles in the upper extremities: biceps brachii, triceps brachii, supinators

27. Which group is incorrect?
    a. contractile proteins: actin, myosin
    b. types of connective tissues: tendons, perimysium, aponeurosis
    c. movements: flexion, supination, abduction
    d. types of involuntary muscles: skeletal, smooth, cardiac

28. Which group is incorrect?
    a. muscles in the upper part of the body: deltoid, biceps brachii, hamstrings
    b. movements: flexion, supination, abduction
    c. muscles in the lower extremities: hamstrings, quadriceps femoris, gastrocnemius
    d. contractile proteins: actin, myosin

CASE STUDY

B.J. was scheduled for abdominal surgery. In addition to the general anesthetic, B.J. also given a curare-like drug during surgery.

1. What was the purpose of the curare-like drug?
    a. to cause sedation
    b. to deaden or numb the entire abdomen
    c. to cause muscle relaxation
    d. to cause unconsciousness

2. Why may the curare-like drug affect respirations?
    a. It depresses the respiratory center in the brain.
    b. It binds or inactivates oxygen.
    c. It interferes with activity in the neuromuscular junction, thereby impairing contraction of the respiratory muscles.
    d. It causes swelling of the tongue and cuts off the flow of air into the lungs.

PUZZLE

Hint: Wore a Cast ... Shrunk My Leg!

Directions: Perform the following functions on the Sequence of Words below. When all the functions have been performed, you are left with a word or words related to the hint. Record your answer below.

Functions: Remove the following:

1. Muscle group on the anterior thigh; muscle group on the posterior thigh

2. Tendon that attaches the gastrocnemius to the calcaneus

3. Muscle that causes plantar flexion

4. Transmitter within the neuromuscular junction

5. Muscles (two) of mastication

6. Muscle terms that mean *round, triangular, short,* and *diagonal*

7. Muscle that attaches to the temporal bone, sternum, and clavicle

8. Muscles that lie along the anterior upper arm and posterior upper arm

9. From Z line to Z line

10. Muscle response to death

11. Muscle response to overuse as in "pumping iron"

12. Storage "envelope" for Ca^{2+} in a relaxed muscle

13. Sustained muscle contraction

14. Breathing muscles (2)

15. Flat sheetlike fascia that attaches muscle to muscle or muscle to bone

Sequence of Words

DELTOIDAPONEUROSISMASSETERSTERNO
CLEIDOMASTOIDSARCOMEREGASTROCNE
MIUSOBLIQUEQUADRICEPSFEMORISDIA
PHRAGMACETYLCHOLINERIGORMORTISACHIL
LESTENDONDISUSEATROPHYTRICEPSBRACHI
ISARCOPLASMICRETICULUMBICEPSBRACHII
HAMSTRINGSINTERCOSTALSBREVISTETA
NUSHYPERTROPHYTEMPORALISTERES

Answer: _____

10 Nervous System: Nervous Tissue and Brain

OBJECTIVES

1. Define the two divisions of the nervous system.
2. List three general functions of the nervous system.
3. Discuss the cellular composition of the nervous system, including:
   - Compare the structure and functions of the neuroglia and neuron.
   - Explain the function of the myelin sheath.
   - Explain how a neuron transmits information.
4. Describe the structure and function of a synapse.
5. Describe the functions of the four major areas of the brain and the four lobes of the cerebrum.
6. Describe how the skull, meninges, cerebrospinal fluid, and blood-brain barrier protect the central nervous system.

Part I: Mastering the Basics

MATCHING

Organization of the Nervous System

Directions: Match the following terms to the most appropriate definition by writing the correct letter in the space provided. Some terms may be used more than once. See text, pp. 173, 174.

A. central nervous system (CNS)
B. peripheral nervous system
C. motor nerves
D. sensory nerves
E. integrative function

1. _____ Nerves that carry out the plans made by the CNS

2. _____ Part of the nervous system that contains the brain and spinal cord

3. _____ Nerves that gather information from the environment and carry it to the CNS

4. _____ Part of the nervous system consisting of sensory and motor nerves that connect the brain and the spinal cord with the rest of the body

5. _____ The processing and interpretation of information by the cells of the CNS; the decision-making capability

MATCHING

Nerve Cells

Directions: Match the following terms to the most appropriate definition by writing the correct letter in the space provided. See text, pp. 174-177.

A. neurons
B. astrocyte
C. ependymal cell
D. Schwann cells
E. ganglia
F. nuclei
G. microglia
H. neuroglia (glia)

1. _____ Common type of glial cell that supports and protects the neurons; helps form the blood-brain barrier

2. _____ Nerve tissue that is called *nerve glue*; composed of astrocytes, microglia, oligodendrocytes, and ependymal cells

3. _____ Nerve cells that transmit information as electrical signals

4. _____ Type of glial cell that lines the inside cavities of the brain; helps form the cerebrospinal fluid (CSF)

5. _____ Glial cells that engage in phagocytosis of pathogens and damaged tissue

6. _____ Glial cells that form the myelin sheath in the peripheral nervous system

7. _____ Clusters of cell bodies located within the CNS

8. _____ Clusters of cell bodies located in the peripheral nervous system

MATCHING

Parts of a Neuron

Directions: Match the following terms to the most appropriate definition by writing the correct letter in the space provided. Some terms may be used more than once. See text, pp. 174-177, 180.

A. cell body
B. axon
C. axon terminal
D. nodes of Ranvier
E. neurilemma
F. myelin sheath
G. dendrites
H. acetylcholine (ACh)

1. _____ Part of the neuron that transmits information away from the cell body

2. _____ Part of the neuron that contains the nucleus; dendrites bring information to this structure, and the axon carries information away from this structure.

3. _____ White fatty material that surrounds the axon; increases the rate at which the electrical signal travels along the axon

4. _____ Short segments of the axonal membrane that are not covered by myelin sheath

5. _____ Part of the axon where the neurotransmitters are stored

6. _____ Treelike part of the neuron that receives information from another neuron; transmits that information toward the cell body

7. _____ Layer that covers the axons of peripheral neurons; concerned with nerve regeneration

8. _____ Makes white matter white

9. _____ Allows for saltatory conduction of the nerve impulse

10. _____ A common neurotransmitter

READ THE DIAGRAM

Two Neurons and a Synapse

Directions: Referring to the illustration, write the numbers in the blanks. See text, p. 182.

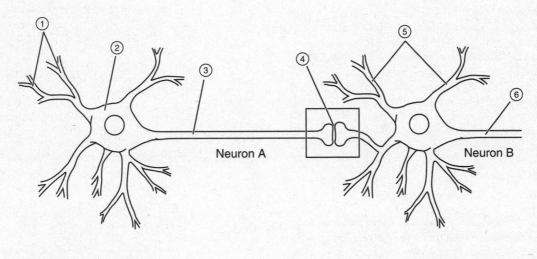

Neuron A   Neuron B

1. _____ The cell body of neuron A

2. _____ The axon of neuron B

3. _____ The synapse

4. _____ The dendrites of neuron B

5. _____ The axon of neuron A

6. _____ The dendrites of neuron A

7. _____ Structure that carries the electrical signal to the cell body of neuron A

8. _____ Point at which neuron A communicates with neuron B

9. _____ Structure that carries the action potential away from the cell body of neuron B

10. _____ and _____ are dendrites.

11. _____ and _____ are axons.

COLORING AND DRAWING

Directions: On the figure above, color the appropriate areas as indicated below.

1. Color the dendrites of neuron A **red.**

2. Color the axon of neuron B **yellow.**

3. Color the presynaptic membrane **blue.**

4. Color the postsynaptic membrane **green.**

DRAW IN

Directions: For the figure on p. 68, draw in the structures indicated below.

1. The storage of the neurotransmitter within the vesicles of the axon terminal

2. The myelin sheath and nodes of Ranvier on the axon of neuron A

3. An arrow that indicates the direction of movement of the nerve impulse (action potential)

MATCHING

Nerve Impulse

Directions: Match the following terms to the most appropriate definition by writing the correct letter in the space provided. Some terms may be used more than once. See text, pp. 175-178.

A. depolarization
B. resting membrane potential (RMP)
C. action potential
D. repolarization

1. _____ This phase of the action potential occurs when the threshold potential has been attained.

2. _____ The inside of the unstimulated neuron is negative; electrical charge is caused by the outward leak of potassium (K^+).

3. _____ The depolarized neuron returns to the resting state.

4. _____ The first phase of the action potential caused by an inward movement of sodium (Na^+)

5. _____ The changes in electrical charge across the membrane during depolarization and repolarization; also called the *nerve impulse*

6. _____ The second phase of the action potential that is caused by the outward movement of potassium (K^+)

MATCHING

Bumps and Grooves

Directions: Match the following terms to the most appropriate definition by writing the correct letter in the space provided. Some terms may be used more than once. See text, pp. 184-185.

A. central sulcus
B. sulcus
C. fissure
D. lateral sulcus
E. gyrus
F. precentral gyrus
G. postcentral gyrus
H. longitudinal fissure

1. _____ Convolution located on the frontal lobe immediately anterior to the central sulcus

2. _____ A bump or elevation on the surface of the cerebrum

3. _____ A shallow groove found on the surface of the brain

4. _____ A deep groove found on the surface of the brain

5. _____ Also called a *convolution*

6. _____ Convolution that is located on the parietal lobe immediately posterior to the central sulcus

7. _____ Sulcus that separates the frontal lobe from the parietal lobes

8. _____ Groove that separates the temporal lobe from the frontal and parietal lobes

9. _____ Deep groove that separates the left and right hemispheres

10. _____ Sulcus that separates the primary motor cortex from the primary somatosensory cortex

MATCHING

Parts of the Brain

Directions: Match the following terms to the most appropriate definition by writing the correct letter in the space provided. Some terms may be used more than once. See text, pp. 183-191.

A. cerebrum
B. brain stem
C. diencephalon
D. pons
E. cerebellum
F. frontal lobe
G. parietal lobe
H. temporal lobe
I. limbic system
J. reticular formation
K. occipital lobe
L. hypothalamus
M. thalamus
N. medulla oblongata
O. corpus callosum

1. _____ Largest part of the brain

2. _____ Called the *emotional brain*

3. _____ We first become aware of pain at this level of the diencephalon; however, this structure does not allow us to determine the type of pain or locate the source of the pain.

4. _____ Part of the brain stem that connects the brain to the spinal cord

5. _____ The central sulcus separates the frontal lobe from this lobe.

6. _____ Composed of the midbrain, pons, and medulla oblongata

7. _____ Part of the diencephalon that controls the pituitary gland; also helps control the autonomic nervous system, water balance, and body temperature

8. _____ Part of the brain that is divided into the right and left hemispheres

9. _____ Composed of the thalamus and the hypothalamus

10. _____ This structure means *bridge;* it helps regulate breathing rate and rhythm.

11. _____ Bands of white matter that join the right and left cerebral hemispheres

12. _____ The precentral gyrus of this cerebral lobe is the major motor cortex; nerve impulses that originate in the motor area control voluntary muscle activity.

13. _____ The postcentral gyrus of this cerebral lobe is the primary somatosensory area.

14. _____ Plays a key role in personality development, emotional and behavioral expression, and performance of high-level thinking and learning tasks

15. _____ Cerebral lobe that contains the primary auditory cortex (hearing) and the olfactory cortex (smell)

16. _____ Part of the brain stem called the *vital center* because it regulates vital processes such as blood pressure, heart rate, and respirations

17. _____ Part of the brain stem that contains the vomiting center

18. _____ Cerebral lobe that contains the frontal eye fields

19. _____ Cerebral lobe that controls motor speech

20. _____ Crossing of most motor fibers occurs here

21. _____ A motor homunculus lives in this cerebral lobe.

22. _____ A sensory homunculus lives in this cerebral lobe.

23. _____ Cerebral lobe that is primarily concerned with vision

24. _____ Part of the diencephalon that acts as a relay and sorting station for most sensory fibers

25. _____ This widespread group of cells is concerned with the sleep-wake cycle and consciousness; signals passing from this structure to the cerebral cortex keep us awake.

26. _____ Cerebral lobe that contains Broca's area

27. _____ Damage to this cerebral lobe causes cortical blindness.

28. _____ Part of the brain stem that contains the emetic center

29. _____ Brain structure that protrudes from under the occipital lobe; concerned primarily with the coordination of skeletal muscle activity

30. _____ Damage to this cerebral lobe causes cortical deafness

31. _____ Part of the brain stem that receives information from the chemoreceptor trigger zone (CTZ)

32. _____ Composed of the frontal, parietal, occipital, and temporal lobes

33. _____ Cerebral lobe that functions as the chief executive officer (CEO)

Brain

Directions: Referring to the diagram, fill in the space with the correct numbers. Some numbers may be used more than once. See text, pp. 183-189.

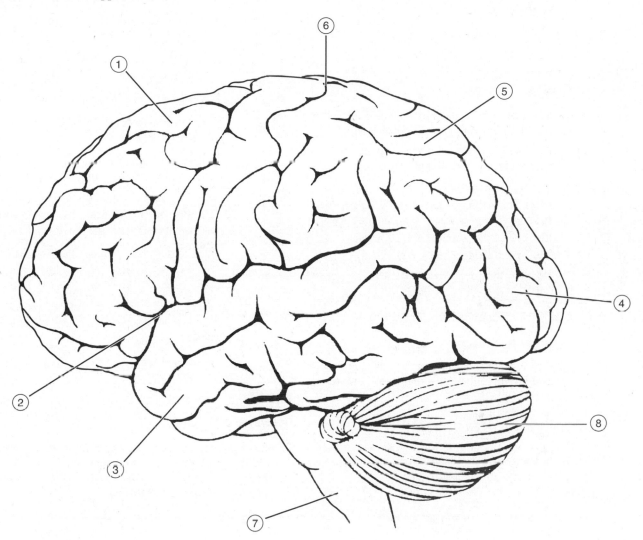

1. _____ Part of the brain stem

2. _____ Called the *vital center* because it regulates respirations, blood pressure, and heart rate

3. _____ Central sulcus that separates the frontal lobe from this lobe

4. _____ The precentral gyrus of this cerebral lobe is the major motor cortex.

5. _____ The postcentral gyrus of this cerebral lobe is the primary somatosensory area.

6. _____ Sulcus that separates the primary motor cortex from the primary somatosensory area

7. _____ Cerebral lobe that is concerned primarily with vision

8. _____ Cerebral lobe that is concerned primarily with hearing

9. _____ The lateral sulcus separates the temporal lobe from this anterior cerebral lobe.

10. _____ This cerebral lobe contains Broca's area.

11. _____ This "little brain" protrudes from under the occipital lobe and is concerned with skeletal muscle coordination.

12. _____ The pyramidal tract arises from this cerebral lobe; the tract carries electrical signals to skeletal muscles, causing voluntary muscle contraction.

13. _____ Destruction of this cerebral lobe causes cortical blindness.

71

14. _____ Destruction of this cerebral lobe causes cortical deafness.

15. _____ This sulcus separates the frontal and parietal lobes from the temporal lobe.

16. _____ Contains the emetic center

17. _____ CEO

COLORING

Xs, Ys, and Zs . . . Yap Yap Yap

Directions: Color or mark the appropriate areas on the illustration on the previous page as indicated below.

1. Color the frontal lobe **red**.

2. Color the occipital lobe **yellow**.

3. Color the parietal lobe **green**.

4. Color the temporal lobe **blue**.

5. Draw a black circle around Wernicke's area.

6. Put a string of *X*s along the precentral gyrus.

7. Put a string of *Y*s along the postcentral gyrus.

8. Put a string of *Z*s along the gyrus that forms a motor homunculus.

9. Label Broca's area as **Yap Yap Yap**.

10. Put an *E* on the frontal eye fields.

MATCHING

Protection of the Brain and Spinal Cord

Directions: Match the following terms to the most appropriate definition by writing the correct letter in the space provided. See text, pp. 190-193.

A. bone
B. blood-brain barrier
C. meninges
D. pia mater
E. dural sinuses
F. ventricles
G. choroid plexus
H. dura mater
I. central canal
J. arachnoid mater
K. subarachnoid
L. arachnoid villi

1. _____ Tough outermost layer of the meninges; means *hard mother*

2. _____ CSF is formed from these blood vessels and ependymal cells that line the ventricular walls.

3. _____ The astrocytes help to form this capillary structure that prevents harmful substances in the blood from diffusing into the brain and spinal cord.

4. _____ Cranium and vertebral column

5. _____ CSF circulates around the brain and spinal cord within this space.

6. _____ Cavities that are filled with blood and help drain the CSF

7. _____ Called the *lateral*, *third*, and *fourth*

8. _____ The soft innermost layer of the meninges; means *soft mother*

9. _____ Finger-like structures that project into the dural sinuses to allow drainage of the CSF

10. _____ Hole in the center of the spinal cord through which CSF flows from the ventricles of the brain to the lower end of the spinal cord

11. _____ The middle layer of the meninges; means *spider* because the layer looks like a spider web

12. _____ Composed of the dura mater, arachnoid mater, and pia mater

13. _____ Meningeal layer that forms the tentorium

ORDERING

Cerebrospinal Fluid

Directions: Trace the formation of CSF from its formation across the choroid plexus in the lateral ventricles to its absorption into the cerebral veins. Use the words listed below. See text, pp. 190-193.

fourth ventricle	arachnoid villi
third ventricle	central canal or foramina
dural sinuses	subarachnoid space

1. choroid plexus of the lateral ventricles

2. _____

3. _____

4. _____

5. _____

6. _____

7. _____

8. cerebral veins

SIMILARS AND DISSIMILARS

Directions: Circle the word in each group that is least similar to the others. Indicate the similarity of the three words on the line below each question.

1. astrocytes   neurons   ependymal   oligodendrocytes cells

_____

2. cell body   dendrites   sulcus   axon

_____

3. neurilemma   myelin sheath   synapse   nodes of Ranvier

_____

4. depolarization   action   reflex   repolarization
                     potential   arc

_____

5. pia   convolution   arachnoid   dura

_____

6. axon   lateral   third   fourth

_____

7. pons   cerebrum   medulla   midbrain
                      oblongata

_____

8. frontal   brain stem   parietal   temporal

_____

9. temporal   parietal   occipital   hypothalamus

_____

10. hypothalamus   central   thalamus   diencephalon
                    sulcus

_____

11. Broca's   occipital   frontal   frontal
    area   lobe   eye fields   lobe

_____

12. acetylcholine   synapse   cerebrospinal   ACh
                               fluid   receptors

_____

13. central   longitudinal   corpus   lateral
    sulcus   fissure   callosum   sulcus

_____

14. Na$^+$ influx   depolarization   pia mater   K$^+$
                                                       efflux

_____

15. serotonin   dopamine   cerebrospinal   GABA
                                 fluid (CSF)

_____

16. gyrus   sulcus   cerebral   convolution
                           "speed bump"

_____

17. temporal   ependyma   hearing   primary
                                     auditory cortex

_____

18. basal   tentorium   shaking   dopamine
    ganglia   palsy   deficient

_____

19. choroid plexus   ependyma   GABA   CSF

_____

20. frontal   vomiting   respiratory   medulla
    lobe   center   control center oblongata

_____

21. motor   foramen   cerebrum   frontal
    homunculus   magnum   lobe

_____

MULTIPLE CHOICE

Directions: Choose the correct answer.

1. A nerve impulse that originates in the precentral gyrus of the cerebrum
   a. allows you to see.
   b. helps you analyze the meaning of speech.
   c. increases respiratory rate.
   d. causes skeletal muscle contraction and movement.

2. The medulla oblongata is called (the)
   a. vital center because it plays an important role in the control of respirations and cardiovascular function.
   b. emotional brain.
   c. primary somatosensory area because it receives information about breathing and heart rate.
   d. Broca's area because it controls motor speech.

3. Which of the following differentiates the functions of the precentral and postcentral gyri?
   a. right and left hemispheres
   b. motor and sensory
   c. neuronal and glial
   d. depolarization and repolarization

4. The central sulcus, lateral fissure, and longitudinal fissure
   a. separate or divide cerebral lobes.
   b. are brain stem structures.
   c. are located only on the right side of the brain.
   d. separate the cerebrum from the cerebellum.

5. The frontal, parietal, occipital, and temporal lobes
   a. are cerebral structures.
   b. comprise the brain stem.
   c. are parts of the diencephalon.
   d. perform only sensory functions.

6. Which of the cerebral lobes is referred to as the somatosensory area?
   a. brain stem
   b. parietal
   c. basal ganglia
   d. precentral gyrus

7. A rapid influx of sodium (Na^+) into a neuron
   a. makes the inside of the neuron more negative than resting membrane potential.
   b. causes repolarization.
   c. causes depolarization.
   d. prevents the firing of an action potential.

8. What is the result of damage to Broca's area?
   a. paralysis of all the extremities
   b. respiratory depression
   c. blindness
   d. impaired motor speech

9. A staggering gait and imbalance are most descriptive of
   a. damage to the occipital lobe.
   b. impaired function of the medulla oblongata.
   c. cerebellar dysfunction.
   d. stimulation of the CTZ.

10. Repolarization of a neuron occurs in response to
    a. a rapid influx of sodium.
    b. a rapid efflux of potassium.
    c. an influx of K^+.
    d. an influx of calcium.

11. The medulla oblongata, pons, and midbrain are
    a. parts of the brain stem.
    b. cerebral lobes.
    c. dopamine-secreting nuclei.
    d. auditory association areas.

12. What happens at a synapse?
    a. saltatory conduction
    b. formation of CSF
    c. chemical transmission of information
    d. synthesis of myelin sheath

13. The choroid plexus is most concerned with
    a. memory.
    b. formation of CSF.
    c. drainage of CSF.
    d. the integrity of the blood-brain barrier.

14. Saltatory conduction refers to the
    a. flow of the CSF.
    b. "leaping" movement of the nerve impulse along an axon.
    c. pH of the CSF.
    d. blood flow through the brain.

15. Which of the following is most descriptive of ganglia?
    a. cells of the blood-brain barrier
    b. cells that comprise the corpus callosum
    c. clusters of cell bodies
    d. myelinated fibers

16. The meninges
    a. include the pia, arachnoid, and dura maters.
    b. cover the axons of all peripheral nerves.
    c. form the blood-brain barrier.
    d. are mucous membranes.

17. The threshold potential is
    a. the degree of depolarization that must be achieved in order to fire an action potential
    b. a repolarizing event
    c. observed only in glial cells
    d. caused by an efflux of K^+

18. Which of the following is least descriptive of the primary auditory cortex?
    a. hearing
    b. temporal lobe
    c. precentral gyrus
    d. sensory

19. What is the usual cause of impaired mental functioning in older adults?
    a. loss of frontal lobe neurons
    b. deterioration of Broca's area
    c. demyelination of all neurons within the CNS
    d. age-related diseases of the blood vessels such as atherosclerosis

20. With which structure is a motor homunculus associated?
    a. Broca's area
    b. Wernicke's area
    c. limbic system
    d. precentral gyrus

21. This diencephalon structure controls endocrine function because of its anatomical relationship to the pituitary gland.
    a. medulla oblongata
    b. corpus callosum
    c. hypothalamus
    d. cerebellum

22. What classification of drug is used to "quiet" the CTZ?
    a. antibiotic
    b. antiemetic
    c. skeletal muscle blocker
    d. anticoagulant

23. Which of the following is most descriptive of cerebral lateralization?
    a. white matter/gray matter
    b. right brain/left brain
    c. sensory/motor
    d. neuronal/glial

24. This cerebral lobe controls motor activity, contains Broca's area, and is the CEO (a reference to its executive function).
    a. frontal
    b. cerebellum
    c. brain stem
    d. corpus callosum

25. Which of the following is the exit point of the brain from the cranium?
    a. central sulcus
    b. foramen magnum
    c. coronal suture
    d. arachnoid villus

CASE STUDY

T.K.O., a professional boxer, sustained a severe blow to his head in round 8. As he left his corner to begin the ninth round, he collapsed to the floor. He was rushed to the emergency department in an unconscious state and was diagnosed with a subdural hematoma. He was placed in a semi-Fowler's position and given a diuretic. Holes were drilled in his skull to relieve the intracranial pressure.

1. Where was the clot located?
   a. within a cerebral ventricle
   b. within the central sulcus
   c. under the outer layer of meninges
   d. within the central canal

2. Which of the following is true about the blood clot?
   a. It may continue to expand as water is pulled into the clot.
   b. It will probably dissolve on its own but generally causes blindness.
   c. It will stop enlarging as soon as the bleeding stops; no treatment is necessary after the bleeding stops.
   d. All blood clots in the brain are lethal, with or without treatment.

3. What caused the loss of consciousness?
   a. The blood clot was pressing on the brain and causing an elevation in intracranial pressure.
   b. The hematoma was producing a brain-toxic substance.
   c. The blood clot was blocking the formation of CSF.
   d. The hematoma caused a brain abscess because a clot is an excellent place for pathogens to grow.

PUZZLE

Hint: The Itsy Bitsy Spider . . . in the CNS

Directions: Perform the following functions on the Sequence of Words that follows. When all the functions have been performed, you are left with a word or words related to the hint. Record your answer in the space provided.

Functions: Remove the following:

1. Hard mother

2. Type of glial cell concerned with the formation of CSF

3. Soft mother

4. The names of the four ventricles

5. CSF circulates within here.

6. The hole in the spinal cord through which CSF flows

7. The cluster of capillaries across which CSF is formed

8. Blood-filled space that drains CSF

9. Cerebral lobes (four)

10. Parts (three) of the brain stem

11. Parts of a neuron (three)

12. Phases (two) of the action potential

13. Band of white matter that connects the right and left cerebral hemispheres

14. Parts of the brain that contain the thalamus and hypothalamus

Sequence of Words

PIAMATERREPOLARIZATIONPONSFRONTAL
EPENDYMAMEDULLAOBLONGATACHO
ROIDPLEXUSPARIETALDURAMATERLATER
ALDENDRITESARACHNOIDMATERCENTRAL
CANALTEMPORALTHIRDCORPUSCALLOSUM
DEPOLARIZATIONOCCIPITALFOURTHCELL
BODYSUBARACHNOIDSPACEAXONDIEN
CEPHALONMIDBRAINDURALSINUS

Answer: _____

11 Nervous System: Spinal Cord and Peripheral Nerves

OBJECTIVES

1. Describe the anatomy of the spinal cord, and list its three functions.
2. Discuss reflexes and list four components of the reflex arc.
3. List and describe the functions of the 12 pairs of cranial nerves.
4. Do the following regarding the peripheral nervous system:
   - Identify the classification of spinal nerves.
   - List the functions of the three major plexuses.
   - Describe a dermatome.
   - Provide the functional classification of the peripheral nervous system.

Part I: Mastering the Basics

MATCHING

Nerve Tracts

Directions: In the spaces provided, indicate whether the following are sensory (S) or motor (M) structures or functions. See text, pp. 201-204.

1. _____ Descending tracts

2. _____ Carries information for touch, pressure, and pain

3. _____ Corticospinal tract

4. _____ Pyramidal tract

5. _____ Ascending tracts

6. _____ Electrical signal arises in the precentral gyrus of the frontal lobe

7. _____ Carries information to the parietal lobe

8. _____ Most neurons decussate in the medulla oblongata

9. _____ Extrapyramidal tracts

10. _____ Spinothalamic tract

11. _____ I feel pain in my little finger.

12. _____ I'm wiggling my toes.

13. _____ I'm cold.

14. _____ I'm winking.

15. _____ I'm hearing voices.

16. _____ Spinocerebellar

17. _____ Afferent fibers

18. _____ Efferent fibers

MATCHING

Reflexes

Directions: Match the following terms to the most appropriate definition by writing the correct letter in the space provided. See text, pp. 204-207.

A. Babinski reflex
B. Achilles tendon reflex
C. gag reflex
D. baroreceptor reflex
E. patellar tendon reflex
F. withdrawal reflex
G. pupillary reflex

1. _____ A protective reflex; quickly moves your finger away from a hot object

2. _____ This reflex helps you maintain a standing posture; also called the *knee-jerk reflex.*

3. _____ This reflex helps your body maintain a normal blood pressure.

4. _____ This reflex is elicited by stroking the sole of the foot; plantar flexion and curling of the toes are normal responses in an adult.

5. _____ This reflex causes the pupils of the eyes to constrict (become smaller) in response to light.

6. _____ A stretch reflex; tapping this tendon in the heel normally causes plantar flexion of the foot; also called the *ankle-jerk reflex.*

7. _____ This reflex involves the glossopharyngeal nerve and helps prevent food and water from going down the wrong way.

Reflex Arc

Directions: Referring to Fig. 11.4 in the textbook, fill in the spaces with the correct numbers. Some numbers may be used more than once. See p. 205.

1. _____ Result of the contraction of the quadriceps femoris

2. _____ Receptors in the thigh muscles are stimulated

3. _____ Motor neuron

4. _____ Sensory neuron

5. _____ Afferent neuron

6. _____ Efferent neuron

7. _____ Extension of the leg

8. _____ Information travels from the spinal cord to the muscle.

9. _____ Information travels from receptors in the muscle to the spinal cord.

MATCHING

Cranial Nerves

Directions: Match the following terms to the most appropriate definition by writing the correct letter in the space provided. Some terms may be used more than once. See text, pp. 207-211.

A. olfactory   F. hypoglossal
B. vestibulocochlear   G. oculomotor
C. vagus   H. trigeminal
D. accessory   I. facial
E. optic

1. _____ Senses hearing and balance

2. _____ The wanderer; widely distributed throughout the thoracic and abdominal cavities

3. _____ Helps control the movements of the tongue; cranial nerve XII

4. _____ Allows you to shrug your shoulders

5. _____ Damage to this nerve causes blindness

6. _____ Sense of smell

7. _____ Tic douloureux, a condition characterized by extreme facial and jaw pain, is caused by inflammation of this nerve.

8. _____ A dilated and fixed pupil is caused by pressure on this nerve.

9. _____ Inflammation of this nerve causes Bell's palsy, a paralysis of one side of the face.

10. _____ Nerve that supplies most of the extrinsic eye muscles; primary function is the movement of the eyeballs

11. _____ Carries sensory information from the retina of the eyes to the occipital lobe of the brain

12. _____ In addition to moving the eyeball, this nerve raises the eyelid and constricts the pupil of the eye.

13. _____ Anosmia

14. _____ Cranial nerve VIII

15. _____ Ototoxicity

16. _____ Ptosis of the lids

17. _____ Cranial nerve II

18. _____ Vertigo

19. _____ Can't smile, wrinkle forehead, secrete tears, or close eyes (on the affected side)

20. _____ Cranial nerve X

MATCHING

Spinal Nerves

Directions: Match the following terms to the most appropriate definition by writing the correct letter in the space provided. Some terms may be used more than once. See text, pp. 210-214.

A. sciatic   E. femoral
B. axillary   F. cauda equina
C. radial   G. phrenic
D. common peroneal   H. plexus(es)

1. _____ Wristdrop is caused by damage to this nerve.

2. _____ Crutch palsy is caused by damage to this nerve.

3. _____ Nerve that supplies the diaphragm, an important breathing muscle

4. _____ Spinal nerves are grouped and sorted here.

5. _____ This large nerve leaves or emerges from the distal end of the spinal cord and supplies the buttocks and posterior thighs.

6. _____ Nerve groupings that are described as cervical, brachial, and lumbosacral nerves

7. _____ Severing of this nerve requires the use of a ventilator

8. _____ Group of nerves that emerge from the distal end of the spinal cord; horse's tail

9. _____ Innervates the inner thigh area

10. _____ If damaged, causes footdrop

11. _____ Must administer an intramuscular injection in the upper outer quadrant of the buttocks to avoid injuring this nerve

SIMILARS AND DISSIMILARS

Directions: Circle the word in each group that is least similar to the others. Indicate the similarity of the three words on the line below each question.

1. descending   sensory   corticospinal   pyramidal

_____

2. ascending   motor   spinothalamic   sensory

_____

3. motor   efferent   descending   spinothalamic

_____

4. spinal nerves   12 pairs   mixed nerves   31 pairs

_____

5. phrenic   diaphragm   motor   gag reflex

_____

6. CN VIII   hearing   vestibulocochlear   facial nerve

_____

7. CN II   blindness   ptosis of the lid   optic

_____

8. CN I   CN III   CN VIII   CN II

_____

9. optic   sciatic   olfactory   oculomotor

_____

10. ulnar   dermatome   radial   median

_____

11. vagus   ptosis of the lid   CN X   "wanderer" nerve

_____

12. cervical   reflex arc   thoracic   lumbar

_____

13. vagus   sciatic   common peroneal   femoral

_____

14. foramen magnum   cervical   lumbosacral   brachial

_____

15. anosmia   footdrop   wristdrop   crutch palsy

_____

16. CN I   vision   motor   olfactory

_____

17. CN III   aphasia   ptosis of the lid   fixed-dilated pupil

_____

18. hemi-   meningo-   para-   quadra-

_____

19. CN VII   optic nerve   "weakest blink"   orbicularis oculi

_____

20. CN IX   glossopharyngeal   Babinski   gag reflex

_____

Part II: Putting It All Together

MULTIPLE CHOICE

Directions: Choose the correct answer.

1. Which of the following is most descriptive of a descending tract?
   a. afferent
   b. sensory
   c. spinothalamic
   d. motor

2. Which of the following is most likely to experience ototoxicity?
   a. a furniture mover who strained his back
   b. a person who was diagnosed with a tumor involving the second cranial nerve
   c. a person who took an antibiotic drug that injured CN VIII
   d. a person with Bell's palsy

3. The pyramidal tract is
   a. the major motor tract that originates in the precentral gyrus.
   b. an ascending tract.
   c. a sensory tract.
   d. also called the spinothalamic tract.

4. A student nurse is instructed to administer an intramuscular injection in the upper outer quadrant of the buttocks to
   a. prevent ototoxicity.
   b. minimize systemic effects of the drug.
   c. avoid penetration of the subarachnoid space.
   d. avoid injury to the sciatic nerve.

5. Which of the following is a function of the spinal cord?
   a. secretes hormones that regulate blood glucose
   b. is the seat of our emotions
   c. acts as an important reflex center
   d. carries sensory information but not motor information

6. Which of the following is least related to the others?
   a. pyramidal tract
   b. extrapyramidal tract
   c. spinothalamic tract
   d. corticospinal tract

7. What is the purpose of myelination?
   a. increases the speed of the nerve impulse
   b. secretes cerebrospinal fluid
   c. increases the phagocytic activity of the glia
   d. separates neurons from the surrounding glia

8. Which of the following is least descriptive of the vagus nerve?
   a. CN X
   b. distributed throughout the chest and abdomen
   c. inflamed vagus nerve causes Bell's palsy
   d. affects the function of the digestive tract

9. Which of the following is a true statement?
   a. The olfactory nerve is a motor nerve.
   b. The CN II is a sensory nerve.
   c. The phrenic, sciatic, and axillary nerves are cranial nerves.
   d. The vagus nerve is confined to the cranium.

10. Which of the following is most descriptive of the cauda equina?
    a. spinal nerves that emerge from the tail end of the spinal cord
    b. cells that secrete cerebrospinal fluid
    c. glial cells that form the blood-brain barrier
    d. meninges

11. Diagnostically, a needle is inserted between the third and fourth lumbar vertebrae into the subarachnoid space to
    a. relieve intracranial pressure from a closed head injury.
    b. obtain a sample of cerebrospinal fluid.
    c. administer blood.
    d. assess the withdrawal reflex.

12. These nerves supply voluntary skeletal muscles, causing movement.
    a. somatic motor nerves
    b. CNs I, II, VIII
    c. optic nerve
    d. vestibulocochlear nerve

13. A mixed nerve is one that
    a. only transmits information for pain.
    b. only transmits information that originates in the precentral gyrus.
    c. contains both sensory and motor fibers.
    d. only affects organs that are in the abdominal cavity.

14. Which involuntary response to a stimulus is accomplished by these four structures: receptor, sensory neuron, motor neuron, effector organ?
    a. action potential
    b. decussation
    c. reflex arc
    d. saltatory conduction

15. What is the effector organ in the knee-jerk or patellar tendon reflex?
    a. quadriceps tendon
    b. quadriceps femoris muscle
    c. spinal cord
    d. gastrocnemius

16. Which of the following is least descriptive of the oculomotor nerve?
    a. CN III
    b. controls the movement of the eyeball
    c. increased intracranial pressure compresses this nerve; causes ptosis of the eyelid
    d. carries sensory information from the eye to the occipital lobe (vision)

17. Which of the following is a consequence of damage to the glossopharyngeal nerve?
    a. inability to shrug the shoulders and move the upper extremities
    b. blindness
    c. loss of the gag reflex and aspiration of food or water into the lungs
    d. loss of balance

18. The phrenic nerve
    a. is a cranial nerve.
    b. exits the spinal cord at the level of T12.
    c. innervates the major breathing muscle.
    d. is classified exclusively as ascending and sensory.

19. The first three cranial nerves
    a. are all sensory.
    b. innervate the eye.
    c. are all motor.
    d. are the olfactory, optic, and oculomotor nerves.

20. Which of the following is true of the spinothalamic tract?
    a. It is a descending tract.
    b. It is also called the pyramidal tract.
    c. It carries the somatic motor neurons.
    d. It is an ascending tract that carries information about temperature, pain, touch, and pressure.

21. Which of the following is least descriptive of the cauda equina?
    a. spinal nerves
    b. brachial plexus
    c. distal spinal cord
    d. innervates lower torso and lower extremities

22. Myel/o refers to the
    a. glial cells that secrete cerebrospinal fluid.
    b. spinal cord.
    c. vertebrae.
    d. herniation of the brain stem.

23. The pyramidal tracts decussate at the medulla oblongata. Which of the following words best describe decussation?
    a. depolarization/repolarization
    b. plexuses (cervical, brachial, lumbosacral)
    c. curvatures (cervical, thoracic, lumbar, sacral)
    d. crossover

24. A person suffers a stroke to the left cerebral hemisphere and suffers a right-sided hemiparalysis. Which of the following words describes the reason for the paralysis of the right side of the body?
    a. cerebral lateralization
    b. anosmia
    c. saltatory conduction
    d. decussation

25. Who likes to have the sole of his foot stroked?
    a. Broca
    b. Cy Attica
    c. Achilles
    d. Babinski

CASE STUDY

Jake and his friends were picnicking near a river. He dove into the river, hitting his head on a submerged rock. When he was pulled from the river by his friends, Jake was conscious but unable to move his body. There was no feeling in his upper or lower extremities. The paramedics stabilized his neck and spinal cord and transported him to the nearest trauma center. He had sustained a fracture at the C6 vertebra.

1. Which of the following is indicated by the paralysis?
   a. The break was accompanied by hemorrhage and severe blood loss.
   b. An infection developed at the fracture site.
   c. The spinal cord had been severed or compressed.
   d. Severe brain damage had occurred.

2. Which of the following words best describe Jake's loss of function?
   a. subdural hematoma
   b. increased intracranial pressure
   c. quadriplegia
   d. poliomyelitis

3. Which statement is true regarding Jake's long-term recovery?
   a. Complete recovery is likely within a 3-month period.
   b. He will regain all motor activity but will not regain any sensory function.
   c. He will require a ventilator to breathe and should regain full use of his upper and lower extremities within 3 months.
   d. It is unlikely that he will regain full use of either his upper or lower extremities.

4. Which statement best explains the reason for the above answer?
   a. Neurons within the CNS do not regenerate.
   b. The reticular activating system reacts to trauma by "closing down"; a deep coma ensues.
   c. Severe injury stops the formation of cerebrospinal fluid.
   d. Injured neurons regenerate but take several months to do so.

81

PUZZLE

Hint: Cleopatra's Favorite Motor Tract

Directions: Perform the following functions on the Sequence of Words that follows. When all the functions have been performed, you are left with a word or words related to the hint. Record your answer in the space provided.

Functions: Remove the following:

1. CN II, sensory, vision

2. Innervates the diaphragm

3. Three nerve plexuses

4. Consequences of severing CNs II and VIII

5. Nerve damaged with crutch palsy

6. Nerves that carry information toward the CNS

7. Nerves that carry information from the CNS toward the effector organs, such as the muscles

8. Mapping of the skin indicating specific innervation

9. Nerve damaged in carpal tunnel syndrome

10. Clinical effects of inflammation of CN VII

11. Damage to this nerve impairs the ability to extend the hip and flex the knee

12. A diagnostic procedure performed by inserting a needle between L3 and L4

13. What reflex is described with stroking of the lateral side of the foot from heel to toe, the toes curl, with slight inversion of the foot

14. CN IX, gag reflex

15. CN VIII, sensory, hearing, balance

16. CN I, smell

17. CN X, "wanderer"

18. CN III, "fixed and dilated," ptosis of the eyelid

19. Another name for the knee-jerk reflex

20. Another name for the ankle-jerk reflex

Sequence of Words

SCIATICDERMATOMEDEAFNESSPATELLAR
TENDONBRACHIALAXILLARYCERVICALVES
TIBULOCOCHLEARPHRENICSENSORYBABIN
SKIPYRAMIDALBELLSPALSYLUMBARPUNC
TUREOPTICMEDIANOCULOMOTORGLOSSO
PHARYNGEALLUMBOSACRALVAGUSBLIND
NESSACHILLESTENDONOLFACTORYCORTI
COSPINALMOTOR

Answer: _____ , _____

BODY TOON

Hint: Another Name for Lumbar Puncture

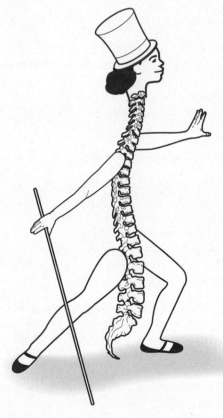

Chapter **11** Nervous System: Spinal Cord and Peripheral Nerves

12 Autonomic Nervous System

Answer Key: Textbook page references are provided as a guide for answering these questions. A complete answer key was provided for your instructor.

OBJECTIVES

1. Describe the function and pathway of autonomic (visceral) reflexes.
2. Do the following regarding the autonomic nervous system:
   - Describe the function of the autonomic nervous system.
   - Identify the two divisions of the autonomic nervous system.
   - State the anatomical and functional differences between the sympathetic and parasympathetic nervous systems.
   - Define autonomic terminology used in pharmacology.
   - Differentiate between autonomic tone and vasomotor tone.
3. Discuss autonomic nervous system neurons, including:
   - Define *cholinergic* and *adrenergic* fibers.
   - Name the major neurotransmitters of the autonomic nervous system.
   - Name and locate the cholinergic and adrenergic receptors.
4. Explain the terms used to describe the effects of neurotransmitters and drugs on autonomic receptors.

Part I: Mastering the Basics

MATCHING

Autonomic Nervous System

Directions: In the spaces provided, indicate whether the following describe sympathetic (S) or parasympathetic (P) nervous system effects. See text, pp. 220-221.

1. _____ Thoracolumbar outflow
2. _____ Feed-and-breed
3. _____ Paravertebral ganglia
4. _____ Stressed-out and uptight
5. _____ Craniosacral outflow
6. _____ Paradoxical fear, "bradying down"
7. _____ Adrenergic
8. _____ Vasomotor tone
9. _____ Fight-or-flight

READ THE DIAGRAM

Directions: Refer to Fig. 12.2 in the textbook, and indicate if the statement is true (T) or false (F). See text, p. 223.

1. _____ The paravertebral ganglia are present in both the parasympathetic nervous system (PNS) and sympathetic nervous system (SNS).
2. _____ The preganglionic fibers of the PNS and SNS are cholinergic.
3. _____ The postganglionic fibers of the PNS and SNS are cholinergic.
4. _____ The postganglionic fibers of the PNS and SNS are adrenergic.
5. _____ The preganglionic fibers of the SNS exit the spinal cord at the thoracolumbar region.
6. _____ The transmitter for the postganglionic fibers of the SNS is norepinephrine (NE).
7. _____ The transmitter for the postganglionic fibers of the PNS is acetylcholine (ACh).
8. _____ The preganglionic and postganglionic fibers associated with the craniosacral outflow are cholinergic.
9. _____ The transmitter of the preganglionic fibers of both the PNS and SNS is ACh.
10. _____ Fibers colored green are cholinergic.
11. _____ Fibers colored red are adrenergic.
12. _____ All autonomic fibers are colored red.
13. _____ Preganglionic fibers of the PNS are longer than the preganglionic fibers of the SNS.
14. _____ Drugs that block the effects of ACh affect both the SNS and PNS.
15. _____ Drugs that block the effects of NE affect both the SNS and PNS.

Autonomic Receptors

Directions: Refer to Fig. 12.3 in the textbook, and indicate if the statement is true (T) or false (F). See text, p. 226.

1. _____ The postganglionic receptors for the PNS are muscarinic.

2. _____ Muscarinic receptors are activated by NE.

3. _____ Alpha (α) and beta (β) receptors are activated by NE.

4. _____ The postganglionic receptors for the SNS are called alpha and beta receptors.

5. _____ Fibers that are colored green secrete ACh as their transmitter.

6. _____ Fibers that are colored red secrete NE as their transmitter.

7. _____ Nicotinic receptors are found only in the autonomic nervous system.

8. _____ The transmitter of the preganglionic fibers of both the PNS and SNS activates nicotinic receptors.

9. _____ A drug that blocks muscarinic receptors diminishes the parasympathetic response.

10. _____ A drug that blocks alpha or beta receptors diminishes a sympathetic response.

11. _____ Muscarinic and nicotinic receptors are cholinergic receptors.

12. _____ Alpha and beta receptors are adrenergic receptors.

13. _____ Cholinergic fibers are found only in the ANS.

14. _____ A drug that affects the N_N receptors affects both the PNS and SNS.

15. _____ Nicotinic receptors within the neuromuscular junction are activated by ACh.

16. _____ Nicotinic receptors are located in both the PNS and SNS.

17. _____ A drug that blocks the N_M receptors within the neuromuscular junction causes skeletal muscle paralysis.

DRAW IT

Receptor Shapes and Fit

Directions: Refer to Fig. 12.3 in the textbook, and draw the following shapes.

1. Draw the shape of the muscarinic receptor. Draw the shape of ACh that fits into this receptor.

2. Draw the shape of the alpha or beta receptor. Draw the shape of the NE that fits into this receptor.

3. Draw the shape of the nicotinic N_M receptor. Draw the shape of the ACh that fits into the receptor.

MATCHING

Sympathetic or Parasympathetic Effects

Directions: Indicate if the following is a sympathetic effect (S) or a parasympathetic effect (P). See text, pp. 220-221, 225-227.

1. _____ Increased heart rate

2. _____ Dilation of the pupils of the eyes

3. _____ Dilation of the breathing passages

4. _____ Stimulation of urination

5. _____ Increased blood pressure

6. _____ Vasoconstriction

7. _____ Decreased heart rate

8. _____ Stronger heart muscle contraction

9. _____ Constriction of the pupil of the eye

10. _____ Increased perspiration

11. _____ Pounding heart and sweaty palms

12. _____ Anxiety and tremors

13. _____ Vagal discharge

SIMILARS AND DISSIMILARS

Directions: Circle the word in each group that is least similar to the others. Indicate the similarity of the three words on the line below each question.

1. feed-and-breed   parasympathetic

   thoracolumbar   resting and digesting

2. cholinergic   norepinephrine

   parasympathetic   craniosacral

_____

3. thoracolumbar   sympathetic

   muscarinic   fight-or-flight

_____

4. alpha receptors   beta receptors

   adrenergic fibers   ACh

_____

5. norepinephrine   cholinergic

   muscarinic   nicotinic

_____

6. craniosacral   sympathetic   cholinergic   ACh

_____

7. N_M N_N muscarinic   alpha, beta

_____

8. muscarinic   norepinephrine   adrenergic   alpha, beta

_____

Part II: Putting It All Together

MULTIPLE CHOICE

Directions: Choose the correct answer.

1. Sympathetic and parasympathetic nerves
   a. are somatic motor neurons.
   b. supply the voluntary skeletal muscles.
   c. include the phrenic, sciatic, and brachial nerves.
   d. innervate the viscera.

2. Which of the following is least true of the SNS?
   a. fight-or-flight
   b. Preganglionic fibers are cholinergic.
   c. Postganglionic fibers are adrenergic.
   d. The postganglionic receptor is muscarinic.

3. Which of the following is most descriptive of the PNS?
   a. fight-or-flight
   b. Preganglionic fibers are adrenergic.
   c. Postganglionic fibers are adrenergic.
   d. The postganglionic receptor is muscarinic.

4. Stimulation of the SNS causes the heart to beat stronger and faster. A drug that also causes the heart to beat stronger and faster is described as
   a. parasympatholytic.
   b. vagomimetic.
   c. sympathomimetic.
   d. sympatholytic.

5. Vasomotor tone is
   a. a vasoconstrictor effect caused by background firing of the sympathetic nerves.
   b. a vagally induced vasoconstriction.
   c. a response to activation of the muscarinic receptors on the blood vessels.
   d. caused by a beta-adrenergic antagonist.

6. What is the clinical consequence of loss of vasomotor tone?
   a. urticaria and pruritus
   b. lethargy and jaundice
   c. severe decline in blood pressure and shock
   d. elevation in blood pressure and hemorrhage

7. Paravertebral ganglia
   a. contain beta-adrenergic receptors that are activated by NE.
   b. are part of the craniosacral outflow.
   c. are located within the SNS.
   d. are located within the effector organs.

8. Which of the following is least descriptive of the thoracolumbar outflow?
   a. fight-or-flight
   b. muscarinic and nicotinic receptors
   c. SNS
   d. paravertebral ganglia

9. Alpha- and beta-adrenergic receptors are
   a. associated with the PNS.
   b. associated with craniosacral outflow.
   c. located on the paravertebral ganglia.
   d. activated by NE.

10. The adrenal medulla secretes epinephrine and NE; the effects of the hormones are best described as
    a. vagolytic.
    b. sympathomimetic.
    c. parasympathomimetic.
    d. sympatholytic.

11. Which of the following fibers secretes NE?
    a. preganglionic sympathetic
    b. preganglionic parasympathetic
    c. postganglionic sympathetic
    d. postganglionic parasympathetic

12. Which of the following is least descriptive of mono-amine oxidase (MAO)?
    a. enzyme that degrades NE
    b. found within all cholinergic fibers
    c. associated with sympathetic activity
    d. associated with adrenergic fibers

13. A beta$_1$-adrenergic agonist
    a. increases heart rate.
    b. causes the release of acetylcholine.
    c. blocks the effects of NE at its receptor site.
    d. lowers blood pressure.

14. Atropine is classified as a muscarinic blocker and therefore is
    a. parasympathomimetic.
    b. sympatholytic.
    c. vagolytic.
    d. sympathomimetic.

15. Muscarinic receptors are located on
    a. the paravertebral ganglia.
    b. the effector organs—postganglionic parasympathetic.
    c. the effector organs—postganglionic sympathetic.
    d. all autonomic ganglia.

16. SNS stimulation causes vasoconstriction of the blood vessels, thereby elevating blood pressure. Which of the following drugs lowers blood pressure?
    a. vagolytic
    b. sympathomimetic
    c. alpha$_1$-adrenergic blocker
    d. beta$_2$-adrenergic agonist

17. SNS stimulation causes relaxation of the breathing passages (i.e., bronchodilation). Which of the following drugs achieves this effect?
    a. beta$_2$-adrenergic agonist
    b. alpha$_1$ blocker
    c. muscarinic antagonist
    d. vagomimetic

18. A patient received an antimuscarinic drug (atropine) preoperatively. What drug-related postoperative consequence is he or she likely to experience?
    a. slow heart rate
    b. inability to urinate
    c. excess salivation
    d. pinpoint pupils

19. Which of the following neurotransmitters activates muscarinic receptors?
    a. acetylcholinesterase
    b. dopamine
    c. NE
    d. ACh

20. Which of the following is least descriptive of cranio-sacral outflow?
    a. parasympathetic
    b. muscarinic and nicotinic receptors
    c. NE
    d. feed and breed

CASE STUDY

A patient has had a heart attack and is experiencing a very slow heart rate because of intense parasympathetic discharge.

1. Which of the following drugs will increase his heart rate?
    a. antimuscarinic
    b. beta$_1$-adrenergic blocker
    c. beta$_1$-adrenergic antagonist
    d. vagomimetic

2. Restate the answer in question 1.
    a. sympatholytic
    b. vagolytic
    c. parasympathomimetic
    d. muscarinic agonist

3. What would happen if a muscarinic agonist were administered?
    a. The heart rate would increase to normal.
    b. The heart rate would become too rapid.
    c. Breathing would cease.
    d. The heart rate would decrease further.

PUZZLE

Hint: An Autonomic Blood Pressure Event

Directions: Perform the following functions on the Sequence of Words that follows. When all the functions have been performed, you are left with a word or words related to the hint. Record your answer in the space provided.

Functions: Remove the following:

1. Neurotransmitter for cholinergic fibers

2. An "outflow" name for the SNS

3. Name of the sympathetic ganglia that run parallel to the spinal cord

4. Neurotransmitter for adrenergic fibers

5. Name of a cholinergic receptor

6. An "outflow" name for the PNS

7. Name of the parasympathetic nerve that innervates the heart

8. Name of two adrenergic receptors

9. Name of a drug that is classified as a beta$_1$-adrenergic blocker

10. Name of a drug that is classified as a muscarinic blocker (anticholinergic)

11. Another name for the thoracolumbar outflow.

12. Another name for the craniosacral outflow.

Sequence of Words

PARAVERTEBRALALPHAPROPRANOLOLBETA
FIGHTORFLIGHTNOREPINEPHRINEVAGUSFEED
ANDBREEDACETYLCHOLINEBARORECEPTOR
REFLEXMUSCARINICATROPINETHORACOLUM
BARCRANIOSACRAL

Answer: _____

13 | Sensory System

OBJECTIVES

1. State the functions of the sensory system.
2. Define the five types of sensory receptors.
3. Describe the four components involved in the perception of a sensation and two important characteristics of sensation.
4. Describe the five general senses.
5. Describe the special senses of smell and taste.
6. Describe the sense of sight, including:
   - Describe the structure of the eye.
   - Explain the movement of the eyes.
   - Describe how the size of the pupils change.
7. Describe the sense of hearing, including:
   - Describe the three divisions of the ear.
   - Describe the functions of the parts of the ear involved in hearing.
   - Explain the role of the ear in maintaining the body's equilibrium.

Part I: Mastering the Basics

MATCHING

Receptors

Directions: Match the following terms to the most appropriate definition by writing the correct letter in the space provided. See text, pp. 232-234.

A. receptor   D. thermoreceptors
B. photoreceptors   E. nociceptors
C. chemoreceptors   F. mechanoreceptors

1. _____ Receptors stimulated by changes in pressure or movement of body fluids

2. _____ A specialized area of a sensory neuron that detects a specific stimulus

3. _____ Receptors stimulated by chemical substances

4. _____ Receptors stimulated by light

5. _____ Receptors stimulated by tissue damage; also called *pain receptors*

6. _____ Receptors stimulated by changes in temperature

MATCHING

General Senses

Directions: Match the following terms to the most appropriate definition by writing the correct letter in the space provided. Some terms may be used more than once. See text, pp. 234-237.

A. pain   D. pressure
B. temperature   E. proprioception
C. touch

1. _____ Nociceptors for this are free nerve endings that are stimulated by tissue damage (caused by chemicals, ischemia, distention, or distortion).

2. _____ Refers to the sense of orientation or position (allows you to locate a body part without looking at it)

3. _____ Thermoreceptors detect this.

4. _____ Tactile receptors detect this.

5. _____ Receptors for this are located in subcutaneous tissue and in the deep tissue.

6. _____ Sensation you feel if you immerse your hand in boiling water

7. _____ Sensation you feel if a tiny insect crawls along your hairy arm

MATCHING

The Eye

Directions: Match the following terms to the most appropriate definition by writing the correct letter in the space provided. Some terms may be used more than once. See text, pp. 239-243.

A. sclera   J. choroid
B. retina   K. pupil
C. vitreous humor   L. canal of Schlemm
D. iris   M. cornea
E. aqueous humor   N. optic disc
F. lens   O. uvea
G. suspensory ligaments   P. palpebrae
H. ciliary body   Q. canthi
I. conjunctiva

1. _____ Anterior extension of the sclera; avascular structure that allows light to enter the eye

2. _____ Outermost layer or tunic of the posterior eyeball

3. _____ Called the *window of the eye* because it is the first structure through which light enters the eye

4. _____ Colored portion of the eye (e.g., brown eyes, blue eyes)

5. _____ Middle tunic; has a rich supply of blood and nourishes the retina

6. _____ The shape of this structure changes in response to contraction and relaxation of the ciliary muscles; it refracts light waves.

7. _____ Venous sinus that drains aqueous humor

8. _____ Contact lenses are placed on this surface.

9. _____ Layer that extends anteriorly to form the ciliary body and the iris

10. _____ Innermost tunic of the posterior eyeball; nervous tissue that contains the photoreceptors

11. _____ The fluid that helps maintain the shape of the anterior cavity

12. _____ Layer that contains the rods and cones

13. _____ Circular opening in the center of the iris

14. _____ Blind spot

15. _____ Fluid that is formed by the ciliary body and drains through the canal of Schlemm

16. _____ Ciliary muscles attach to these bands of connective tissue that pull on the lens.

17. _____ Mucous membrane that lines the inner surface of the eyelids and folds back to cover a part of the anterior surface of the eyeball

18. _____ Layer that includes the macula lutea and fovea centralis

19. _____ Gel-like substance that fills the posterior cavity

20. _____ Structure that secretes aqueous humor and gives rise to intrinsic eye muscles called the *ciliary muscles*

21. _____ Composed of muscles that determine the size of the pupil

22. _____ Layer that sends information along the optic nerve to the occipital lobe

23. _____ The word that includes the choroid, the ciliary muscle, and the iris

24. _____ Gel-like fluid that gently pushes the retina against the choroid

25. _____ Layer that contains melanocytes to diminish glare as light enters the posterior cavity

26. _____ Corner meeting place for the eyelids

27. _____ Eyelids

MATCHING

Muscles and Nerves of the Eye

Directions: Match the following terms to the most appropriate definition by writing the correct letter in the space provided. Some terms may be used more than once. See text, pp. 243-246.

A. optic nerve
B. ciliary muscle
C. levator palpebrae superioris
D. occipital lobe
E. oculomotor nerve
F. extrinsic eye muscles
G. abducens nerve
H. orbicularis oculi
I. trochlear nerve
J. radial muscle
K. iris
L. facial nerve
M. circular muscle

1. _____ Contains the radial muscle and circular muscles

2. _____ Nerve that carries information from the photoreceptors to the primary visual cortex

3. _____ Cranial nerve (CN) II

4. _____ Sensory nerve for vision

5. _____ Muscle that raises the eyelid

6. _____ Location of the primary visual cortex

7. _____ Elevated intracranial pressure compresses this nerve to cause ptosis of the eyelid.

8. _____ Muscles that move the eyeball in its socket

9. _____ The meaning of 4 in LR_6SO_4

10. _____ The meaning of 6 in LR_6SO_4

11. _____ The structures indicated by LR and SO in LR_6SO_4

12. _____ Elevated intracranial pressure compresses this nerve to cause fixed and dilated pupils.

13. _____ Impaired nerve function that diminishes lacrimation and the ability to close the eye

14. _____ Muscle that contracts to cause a miotic effect

15. _____ Muscle that contracts to cause mydriasis

16. _____ Muscle that contracts and relaxes to change the shape of the lens

17. _____ Damage to this nerve indicates why Mr. Bell has the weakest blink.

18. _____ Muscle that closes the eye

19. _____ Contraction of this muscle widens the pupil.

20. _____ Contains the muscles that cause mydriasis and miosis

21. _____ Includes the superior, inferior, medial, and lateral rectus; also includes the superior and inferior oblique

22. _____ Nerve that controls all extrinsic eye muscles except the lateral rectus and superior oblique

23. _____ Muscles that allow you to look at the ceiling without moving your head

24. _____ Muscle that attaches to the suspensory ligaments

25. _____ Structure that involves the optic chiasma

26. _____ Severing of this nerve causes blindness

27. _____ Nerve that contains only sensory fibers

MATCHING

Disorders of the Eye

Directions: Match the following terms to the most appropriate definition by writing the correct letter in the space provided. Some terms may be used more than once. See text, pp. 239-249. You may also need to refer to the Medical Terminology and Disorders table at the end of the chapter.

A. presbyopia
B. myopia
C. glaucoma
D. detached retina
E. night blindness
F. strabismus
G. conjunctivitis
H. ptosis
I. cataracts
J. choked disc
K. astigmatism
L. aphakia
M. macular degeneration
N. nystagmus
O. retrolental fibroplasia
P. suppression amblyopia
Q. diabetic retinopathy

1. _____ The eyelid is not completely raised, giving the person a sleepy appearance.

2. _____ Absence of the lens when a cataract is extracted surgically

3. _____ An error of refraction described by an older adult who says that his or her arms are getting shorter

4. _____ Consequence of impaired drainage of aqueous humor

5. _____ Deficiency of vitamin A that affects the functioning of the rods and makes it difficult to see in dim light; depletion of rhodopsin

6. _____ Swelling of the optic disc (indication of increased intracranial pressure)

7. _____ An error of refraction usually caused by a flattening or uneven curvature of the cornea

8. _____ Nearsightedness

9. _____ Crossed eyes; classified as convergent or divergent

10. _____ A disease characterized by increased intra-ocular pressure

11. _____ Occurs when the nervous inner layer of the eye falls away from the choroid and is thereby deprived of an adequate blood supply

12. _____ Clouding of the lens that impairs the transmission of light

13. _____ Retinopathy associated with prematurity of the neonate and the administration of high doses of oxygen

14. _____ Formation of microaneurysms/retinal scarring in a person with diabetes mellitus

15. _____ Involuntary and rapid oscillating movements of the eyeball; classified as vertical, horizontal, and rotary or tortional

16. _____ Age-related deterioration of an area of the retina; classified as dry and wet

17. _____ An infection; also called *pinkeye*

18. _____ Loss of vision due to failure to use the eye as a young person

Eyeball

Directions: Referring to the diagram, fill in the spaces with the correct numbers. Some numbers may be used more than once. See text, pp. 241-244.

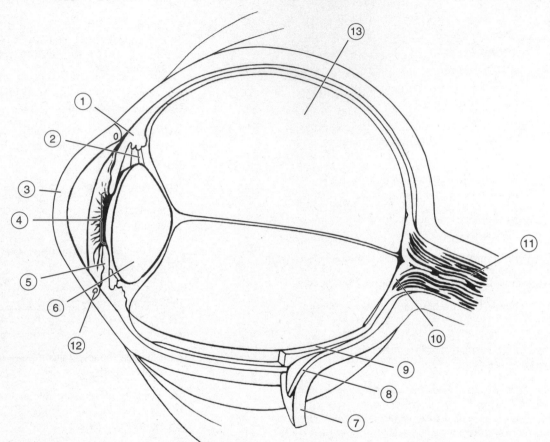

1. _____ Anterior extension of the sclera; avascular structure that allows light to enter the eye

2. _____ Outermost layer or tunic of the eyeball; composed of thick, fibrous connective tissue

3. _____ Called the *window of the eye* because it is the first structure through which light enters the eye

4. _____ Colored portion of the eye (e.g., brown eyes, blue eyes); composed of muscles that determine the size of the pupil

5. _____ Middle tunic; has a rich supply of blood and nourishes the retina

6. _____ The shape of this structure changes in response to the contraction and relaxation of the ciliary muscles; it refracts light waves.

7. _____ Venous sinus that drains aqueous humor

8. _____ Contact lenses are placed on this surface.

9. _____ Layer that extends anteriorly to form the ciliary body and the iris

10. _____ Innermost tunic of the posterior eyeball; the nervous tissue that contains the photo-receptors

11. _____ Optic nerve

12. _____ Clouding of this structure causes cataracts.

13. _____ The circular opening in the center of the iris

14. _____ The blind spot

15. _____ Gel-like substance that fills the posterior cavity

16. _____ Ciliary muscles attach to these bands of connective tissue that pull on the lens.

17. _____ Structure that secretes aqueous humor and gives rise to intrinsic eye muscles called *ciliary muscles*

18. _____ This layer includes the macula lutea and fovea centralis.

19. _____ Impaired drainage through this structure causes glaucoma.

COLORING

Directions: Color the appropriate areas on the illustration on the previous page as indicated below.

1. Color the space occupied by the aqueous humor *red*.

2. Color the space occupied by the vitreous humor *purple*.

3. Color the retina *yellow*.

4. Color the choroid, ciliary body, and iris *green*.

5. Color the sclera and cornea *blue*.

6. Color the lens *orange*.

ORDERING

Directions: Place in order six structures listed below through which light must pass to activate the photoreceptors. Some structures listed below will NOT be used.

Vitreous humor

Ciliary muscle

Sclera

Cornea

Rods and cones

Aqueous humor

Conjunctiva

Lens

Pupil

1. _____

2. _____

3. _____

4. _____

5. _____

6. _____

READ THE DIAGRAM

Directions: Referring to the diagram, fill in the spaces with the correct numbers. Some numbers may be used more than once. See text, pp. 240, 241.

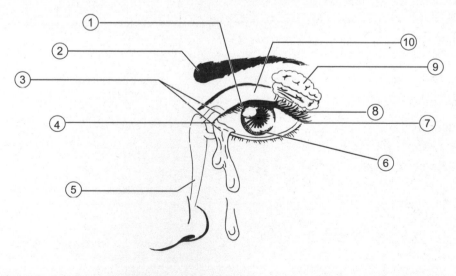

1. _____ Meeting place (closer to the ears) of the eyelids

2. _____ Meeting place of the sclera and the cornea

3. _____ Structure lined with the conjunctiva

4. _____ Patch of hair that is located superior to the eye; traps dust and diminishes glare

5. _____ Secretes tears

6. _____ A gland located in the upper lateral quadrant of the orbit

7. _____ Tears leave the surface through these tiny holes.

8. _____ Pressure on CN III causes ptosis of this structure.

9. _____ Tears flow from the lacrimal puncta into this structure.

10. _____ Damage to the facial nerve causes decreased secretion of this gland.

11. _____ Structure that delivers tears to the nasal cavity

12. _____ Meeting place (closer to the nose) of the eyelids

13. _____ This structure is raised by the levator palpebrae superioris.

14. _____ The junction that is called the limbus

STORY

How to See

Directions: Complete the paragraph by writing the correct term in the space provided. See text, pp. 242-247.

cones   lens
occipital   retina
cornea   pupil
rods   optic
fovea centralis (macula lutea)

Light waves pass through the _____, the avascular extension of the sclera. The light goes through the

opening in the iris, called the _____, and then

through a refracting structure, called the _____.

In daylight, the light waves focus on the _____, the area of most acute vision because of the high numbers

of _____, the photoreceptors for color vision. In dim light, the light waves are scattered along the periphery

of the retina, stimulating the _____, the photoreceptors for night vision. Action potentials (nerve impulses) are formed by the stimulated photoreceptors

located in the inner layer of the eye, called the _____. The nerve impulses travel along the _____ nerve to the primary visual cortex of the _____ lobe of the cerebrum.

MATCHING

Parts of the Ear

Directions: Match the following terms to the most appropriate definition by writing the correct letter in the space provided. Some terms may be used more than once. See text, pp. 249-252.

A. external ear   B. middle ear
C. inner ear

1. _____ Location of the semicircular canals, cochlea, and vestibule

2. _____ Location of the hammer, anvil, and stirrup

3. _____ Ototoxicity (e.g., damage to CN VIII by antibiotics) occurs here.

4. _____ Location of the auditory canal and cerumen

5. _____ The eustachian tube connects the pharynx with this part of the ear.

6. _____ Location of the organ of Corti

7. _____ The tympanic membrane separates the middle ear from this part of the ear.

8. _____ CN VIII originates within this part of the ear.

9. _____ Otitis media occurs here.

10. _____ Location of the malleus, incus, and stapes

11. _____ Ménière's disease occurs here.

12. _____ Vibration of bone occurs here.

13. _____ Location of the labyrinth, perilymph, and endolymph

14. _____ Nerve conduction deafness occurs here.

15. _____ Bone conduction deafness occurs here.

16. _____ Location of the auricle or pinna

17. _____ "Rock and roll" deafness occurs here.

Ear

Directions: Referring to the diagram, place the correct number in the space. Some numbers may be used more than once. See text, pp. 249-252.

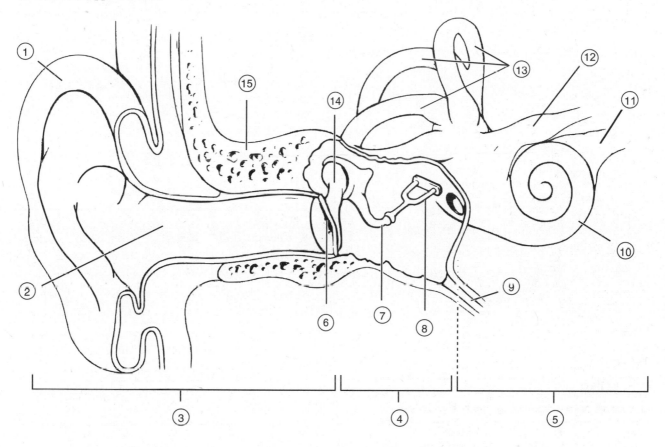

1. _____ Ossicle that sits in the oval window and transmits vibrations to the inner ear

2. _____ Structure that connects the middle ear with the throat

3. _____ Semicircular canals that function in balance or equilibrium

4. _____ Vestibulocochlear nerve

5. _____ Snail-like structure in the inner ear that contains the organ of Corti

6. _____ Ossicle that picks up vibrations from the eardrum

7. _____ Structure that separates the external ear from the middle ear

8. _____ Ossicle located between the malleus and stapes

9. _____ Ossicle that is called the *stirrup* or *stapes*

10. _____ Long tube-like structure that is part of the external ear

11. _____ Otitis media occurs here.

12. _____ "Glue ear" occurs here.

13. _____ Bone that contains the external auditory meatus

14. _____ Ossicle that is called the *malleus* or *hammer*

15. _____ Nerve conduction deafness affects this structure.

16. _____ Hearing is impaired when cerumen is packed against this structure.

COLORING

Directions: Color or draw the appropriate areas on the illustration on the previous page as indicated below.

1. Color the external ear *blue*.

2. Color the middle ear *red*.

3. Color the inner ear *yellow*.

4. Put an *X* on the malleus.

5. Put a *Y* on the incus.

6. Put a *Z* on the stapes.

7. Draw some germs crawling up the eustachian tube into the middle ear.

8. Draw a blob of cerumen leaning up against the tympanic membrane.

9. Place a tube through the tympanic membrane. (Congrats! You have just performed a tympanostomy.)

STORY

How to Hear

Directions: Complete the paragraph by writing the correct term in the space provided. Some terms may be used more than once. See text, pp. 248-251.

endolymph	temporal
incus (anvil)	organ of Corti
cochlear	tympanic membrane
malleus (hammer)	stapes (stirrup)

Sound waves travel through the external auditory canal and bump into the _____, causing the "drum" to vibrate. The vibration is transmitted to the three tiny ossicles called the _____, _____, and _____. The _____ vibrates against the oval window, causing the fluid within the inner ear to move. The inner ear fluid, called the _____, bends the hairlike projections of the receptors, called the _____. When the hairs are bent, action potentials are fired; the nerve impulses travel along the _____ branch of the CN VIII to the primary auditory cortex of the _____ lobe of the cerebrum.

SIMILARS AND DISSIMILARS

Directions: Circle the word in each group that is least similar to the others. Indicate the similarity of the three words on the line below each question.

1. chemo-   noci-   photo-   audio-

   _____

2. pain   touch   vision   proprioception

   _____

3. hearing   pressure   pain   temperature

   _____

4. taste   smell   hearing   proprioception

   _____

5. balance   pain   sight   taste

   _____

6. bitter   salty   hot   sour

   _____

7. lacrimal   lacrimal   gustatory   tears
   glands   puncta

   _____

8. retina   eyebrows   eyelids   palpebrae

   _____

9. sclera   retina   vitreous humor   choroid

   _____

10. retina   rods   conjunctiva   cones

    _____

11. refraction   photoreceptors   lens   emmetropia

    _____

12. hyperopia   refraction   myopia   glaucoma

    _____

13. vitreous   canal of   anterior   aqueous
    humor   Schlemm   chamber   humor

    _____

14. mydriasis   pupil   iris   optic chiasm

_____

15. retina   fovea   sclera   macula
              centralis   lutea

_____

16. CN II   CN III   vagus   optic nerve

_____

17. superior   inferior   radial   medial
    rectus   oblique   rectus

_____

18. circular   radial   iris   lens

_____

19. optic   refraction   CN II   optic
    chiasm   tract

_____

20. astigmatism   myopia   cataracts   farsighted

_____

21. external   middle   inner   temporal
                                   lobe

_____

22. malleus   eustachian tube   incus   stapes

_____

23. hammer   anvil   stirrup   tympanic
                                  membrane

_____

24. cochlea   eustachian   semicircular   vestibule
               tube   canal

_____

25. organ of   eustachian   cochlea   CN VIII
    Corti   tube

_____

26. rods   organ of Corti   photoreceptors   cones

_____

27. balance   semicircular   middle ear   CN VIII
               canals   ossicles

_____

28. endolymph   cerumen   perilymph   inner ear

_____

29. lens   retina   tympanic membrane   cornea

_____

30. glaucoma   cataracts   otitis   macular
                             media   degeneration

_____

31. ototoxicity   Ménière's   glaucoma   otitis
                   disease   media

_____

32. fovea   macula   ptosis of   cones
    centralis   lutea   the eyelid

_____

33. pain   nociceptor   pyramidal   spinothalamic

_____

34. meibomian   tarsal   blind   eyelid
    gland   gland   spot

_____

Part II: Putting It All Together

MULTIPLE CHOICE

Directions: Choose the correct answer.

1. Which of the following is most related to phantom limb pain?
   a. photoreceptor
   b. adaptation
   c. projection
   d. accommodation

2. Which of the following receptors adapt most rapidly?
   a. receptors that detect blood chemistries
   b. olfactory receptors
   c. pain receptors
   d. nociceptors

3. Which of the following is most related to the spino-thalamic tract, thalamus, and nociceptors?
   a. temperature
   b. speech
   c. pain
   d. sight

4. Which of the following is a true statement?
   a. CN I carries information from the retina to the occipital lobe.
   b. The olfactory nerve is motor.
   c. The receptors for the gustatory sense are located on the tongue.
   d. The receptors for gustation are mechanoreceptors.

5. Which of the following is true about tears?
   a. prevent corneal ulceration
   b. form the vitreous humor
   c. form aqueous humor
   d. drain through the canal of Schlemm

6. Which of the following describes the basis for referred pain (e.g., the pain of a heart attack radiates to the left shoulder and arm)?
   a. Myocardial enzymes from the injured heart muscle diffuse into the shoulder region.
   b. The same blood vessels that supply the myocardium also supply the shoulder and arm.
   c. The same nerves that carry sensory information from the heart also carry sensory information from the shoulder.
   d. Nociceptors are found only in the axillary region.

7. Refraction is accomplished when
   a. the pupils constrict.
   b. a sensation is projected back to the receptor.
   c. mydriasis occurs.
   d. light waves are bent.

8. All the extrinsic eye muscles
   a. determine pupillary size.
   b. move the eyeball in the socket.
   c. refract light waves.
   d. change the shape of the lens.

9. Which of the following is not true of the retina?
   a. dependent on the choroid for oxygenation and nourishment
   b. contains the photoreceptors
   c. is the nervous layer of the eye
   d. covers the optic disc, making it the area of most acute vision

10. Myopia, astigmatism, and hyperopia are all
    a. treated by surgically removing the lens.
    b. conditions of farsightedness.
    c. causes of blindness.
    d. errors of refraction.

11. The malleus, incus, and stapes
    a. are inner ear structures.
    b. are located within the semicircular canals.
    c. contain the organ of Corti.
    d. are ossicles located in the middle ear.

12. What happens at the optic chiasm?
    a. Aqueous humor is secreted.
    b. Rods and cones are stimulated.
    c. Vitreous humor is drained from the posterior cavity.
    d. Fibers of the optic nerve of each eye cross and project to the opposite side of the brain.

13. When drainage of the canal of Schlemm is impaired,
    a. lacrimation ceases.
    b. intraocular pressure increases.
    c. the person develops presbyopia.
    d. a cataract forms in the affected eye.

14. Which of the following is most associated with the rods?
    a. color vision
    b. macula lutea
    c. fovea centralis
    d. night vision

15. Which of the following is most related to the sense of hearing?
    a. organ of Corti
    b. chemoreceptors
    c. cranial nerve VII
    d. occipital lobe

16. What is the result of contraction of the radial muscles of the eye?
    a. The eyeball looks toward the sky.
    b. The eyeball looks toward the nose.
    c. mydriasis
    d. The eyelid opens.

17. Which of the following is least related to the middle ear?
    a. bone conduction
    b. eustachian tube
    c. malleus, incus, and stapes
    d. cochlea

18. What causes ototoxicity?
    a. clouding of the lens
    b. paralysis of the ciliary muscles
    c. damage to the cochlear nerve
    d. collection of cerumen in the external ear

19. Which of the following refers to the reflex ability of the lens to change its shape as an object moves closer to the eye?
    a. accommodation
    b. adaptation
    c. projection
    d. presbyopia

20. Because the pituitary gland is located behind the optic chiasm, a pituitary tumor is most likely to cause which condition?
    a. cataracts
    b. a disturbance in vision
    c. error of refraction, such as myopia
    d. conjunctivitis

21. Which of the following is least descriptive of the organ of Corti?
    a. mechanoreceptors
    b. eustachian tube
    c. cochlea
    d. inner ear

22. Which of the following is a balance-related inner ear structure?
    a. semicircular canal
    b. organ of Corti
    c. eustachian tube
    d. cochlear nerve

23. The primary auditory cortex
    a. receives sensory input from CN II.
    b. is located in the temporal lobe.
    c. is called Wernicke's area.
    d. initiates the nerve impulses that move the eyeball.

24. The eustachian tube connects the
    a. throat with the pharynx.
    b. pharynx with the middle ear.
    c. organ of Corti with CN VIII.
    d. stapes with the eustachian tube.

25. What term is used to describe the choroid, ciliary muscle, and iris?
    a. meibomian
    b. uvea
    c. canthus
    d. lacrimal apparatus

26. These glands secrete an oily substance that coats the outer surface of the anterior eyeball and reduces the evaporation of tears.
    a. lacrimal glands
    b. ciliary body
    c. meibomian glands
    d. cerumen-secreting glands

27. Rhodopsin and vitamin A are most associated with
    a. the production of tears by the lacrimal gland
    b. night vision
    c. the secretion of aqueous humor by the ciliary body
    d. refraction

28. Presbyopia and presbycusis both refer to
    a. refractive disorders.
    b. disorders of the eye.
    c. instruments used to assess disorders of the eye and ear.
    d. age-related disorders of the eye and ear.

29. Accommodation, convergence, and pupillary constriction occur
    a. whenever the room lights are dimmed.
    b. when an object moves closer to the eye.
    c. whenever sound is very loud.
    d. whenever the primary visual cortex is activated.

30. A person who is photophobic
    a. is color blind.
    b. has cataracts.
    c. is fearful of pain-inducing exposure to light.
    d. is blind.

CASE STUDY

Six-month-old I.R. awoke at 3 AM crying and pulling on her left ear. There was dried, green, purulent drainage on her earlobe and on the sheet of her crib. Her temperature was 38.3°C (101°F). The pediatrician prescribed an antibiotic and an analgesic.

1. What is the most likely cause of her symptoms?
    a. Ménière's disease
    b. wax in her outer ear that was putting pressure on the eardrum
    c. presbycusis
    d. otitis media (middle ear infection)

2. What is the cause of the purulent drainage?
    a. inner ear fluid
    b. endolymph mixed with ear wax
    c. pus that formed within the middle ear as part of the infectious process
    d. cerumen

3. What is indicated by the presence of purulent drainage on her earlobe?
    a. The infection had run its course and was subsiding.
    b. No antibiotic was necessary.
    c. The tympanic membrane had ruptured.
    d. Fluid from the inner ear was leaking into the external auditory canal.

PUZZLE

Hint: Visually, What's up With Bella Donna?

Directions: Perform the following functions on the Sequence of Words that follows. When all the functions have been performed, you are left with a word or words related to the hint. Record your answer in the space provided.

Functions: Remove the following:

1. Layers (three) of the eyeball

2. Photoreceptors (two)

3. Humors (two)

4. Refracting structure

5. Intrinsic eye muscles (three)

6. Nerve that stimulates most of the extrinsic eye muscles

7. Area of the retina that contains a large number of cones (two)

8. Light penetrates the iris through this hole

9. Part of the eye that contains the radial and circular muscles

10. Drainage point of aqueous humor

11. Cerebral lobe that contains the primary visual cortex

12. The word that includes the choroid, iris, and ciliary muscle

13. Part of the visual pathway where fibers of the optic nerve project to the opposite side of the brain

14. Increased intraocular pressure, "crossed eyes," and cloudy lens

15. Gland that secretes tears

16. The bending of light waves

Sequence of Words

```
O C U L O M O T O R U V E A F O V E A C E N T R
A L I S C A T A R A C T R E T I N A L E N S C I L
I A R Y C O N E S G L A U C O M A P U P I L L A
R Y D I L A T I O N A Q U E O U S H U M O R P U
P I L C I R C U L A R M A C U L A L U T E A O C
C I P I T A L I R I S R E F R A C T I O N M Y D R I
A S I S C H O R O I D R A D I A L S T R A B I S M
U S V I T R E O U S H U M O R L A C R I M A L S
C L E R A O P T I C C H I A S M C A N A L O F S C
H L E M M R O D S
```

Answer: _____, _____

BODY TOON

Hint: Where the Tears Are!

14 Endocrine System

OBJECTIVES

1. List the functions of the endocrine system.
2. Discuss the role and function of hormones in the body, including:
   - Define *hormone*.
   - Explain the process by which hormones bind to the receptor sites of specific tissues (targets).
   - Explain the three mechanisms that control the secretion of hormones.
3. Discuss the pituitary gland, including:
   - Describe the relationship of the hypothalamus to the pituitary gland.
   - Describe the location, regulation, and hormones of the pituitary gland.
4. Identify the other major endocrine glands and their hormones, and explain the effects of hyposecretion and hypersecretion.

Part I: Mastering the Basics

MATCHING

Hormone Action

Directions: Match the following terms to the most appropriate definition by writing the correct letter in the space provided. Some terms may be used more than once. See text, pp. 263-267.

A. target tissue (organ)
B. receptor(s)
C. hormones
D. endocrine glands
E. exocrine glands
F. negative feedback control
G. biorhythm(s)
H. tropic hormones
I. second chemical messenger
J. positive feedback control

1. _____ Types of glands that secrete hormones; called *ductless glands*

2. _____ Types of glands that secrete into ducts (e.g., sweat glands, sebaceous glands)

3. _____ Hormones that are "aimed at" a target

4. _____ Describes this pattern of hormone secretion: cortisol secretion is highest in the morning (8 AM), lowest in the evening (midnight).

5. _____ Endocrine secretions that are classified as proteins, protein-like substances, or steroids

6. _____ Describes a specific tissue or organ to which a hormone binds

7. _____ Describes, for example, this sequence of events: corticotropin-releasing hormone (CRH) stimulates the release of adrenocorticotropic hormone (ACTH), which in turn stimulates the secretion of cortisol; as the blood level of cortisol increases, it shuts off the further secretion of ACTH and CRH.

8. _____ An example is a circadian rhythm.

9. _____ Jet lag and night shift work alter this pattern of hormone secretion.

10. _____ A self-amplification cycle in which a change is the stimulus for an even greater change in the same direction

11. _____ Hormones bind to these special areas in the cell or on the surface of the cell membrane of the target tissue.

12. _____ An example is the menstrual cycle.

13. _____ Cyclic adenosine monophosphate (cAMP)

MATCHING

Glands

Directions: Match the following terms to the most appropriate definition by writing the correct letter in the space provided. Some terms may be used more than once. See text, pp. 268-277.

A. thyroid gland
B. anterior pituitary gland
C. posterior pituitary gland
D. hypothalamus
E. pancreas
F. pineal gland
G. parathyroid glands
H. adrenal cortex
I. adrenal medulla
J. ovaries
K. testes
L. thymus gland

1. _____ Beta and alpha cells of the islets of Langerhans

2. _____ Gonads that secrete estrogen and progesterone

3. _____ Secretes T_3, T_4, and calcitonin

4. _____ Secretes releasing hormones

5. _____ Secretes antidiuretic hormone (ADH) and oxytocin

6. _____ Called the *neurohypophysis*

7. _____ The hypothalamus and this gland are parts of the hypothalamic-hypophyseal portal system.

8. _____ Called the *adenohypophysis*

9. _____ Secretes tropic hormones such as thyroid-stimulating hormone (TSH), ACTH, growth hormone (somatotropic hormone), and the gonadotropins

10. _____ Secretes the catecholamines, epinephrine, and norepinephrine

11. _____ Secretes the steroids—glucocorticoids, mineralocorticoids, and androgens

12. _____ Target gland for ACTH

13. _____ Low plasma levels of calcium stimulate these glands to secrete parathyroid hormone (PTH).

14. _____ Secretes insulin and glucagon

15. _____ Uses iodine to synthesize its hormones

16. _____ Target gland of TSH

17. _____ Secretions of this gland contribute to the sympathetic fight-or-flight response.

18. _____ The gonad that secretes testosterone

19. _____ This gland is associated with these words: *isthmus, follicular cells, colloid,* and *iodine.*

20. _____ This gland plays an important role in immunity and involutes, or shrinks, after puberty.

21. _____ Secretes the hormones that are concerned with sugar, salt, and sex

22. _____ A cone-shaped gland located in the brain; secretes melatonin; called the *biological clock*

23. _____ Gland that enlarges as a goiter

24. _____ Glands that are sometimes embedded within the thyroid gland; removal causes hypocalcemic tetany.

MATCHING

Hormones

Directions: Match the following terms to the most appropriate definition by writing the correct letter in the space provided. Some terms may be used more than once. See text, pp. 268-277.

A. catecholamines H. T_3 and T_4
B. steroids I. calcitonin
C. insulin J. ACTH
D. glucagon K. releasing hormones
E. PTH L. prolactin
F. growth hormone M. ADH
G. gonadotropins N. oxytocin

1. _____ Also called somatotropic hormone, this anterior pituitary hormone stimulates the growth of the musculoskeletal system.

2. _____ The only hormone that lowers blood glucose

3. _____ Cortisol, aldosterone, and testosterone

4. _____ Epinephrine and norepinephrine (NE); sympathomimetic hormones

5. _____ Pancreatic hormone that increases blood glucose levels

6. _____ Hormone that acts on three target organs (bones, kidneys, and intestine) to increase the blood levels of calcium

7. _____ A tropic hormone that is suppressed by increasing plasma levels of cortisol

8. _____ TSH stimulates the thyroid gland to release these hormones.

9. _____ Include follicle-stimulating hormone (FSH) and luteinizing hormone (LH)

10. _____ Stimulates the breasts to make milk; also called *lactogenic hormone*

11. _____ Posterior pituitary hormone that enhances labor

12. _____ Parafollicular cells in the thyroid gland that stimulate osteoblastic activity

13. _____ Iodine-containing hormones that regulate basal metabolic rate (BMR)

14. _____ Posterior pituitary hormone that stimulates the kidneys to reabsorb water

15. _____ Hypothalamic hormones that control the hormonal secretion of the anterior pituitary gland

16. _____ Tropic hormone that stimulates the adrenal cortex to secrete cortisol

17. _____ The stimulus for its release is an increasing blood glucose level.

18. _____ Posterior pituitary hormone that is secreted in response to a low blood volume and concentrated blood, as occurs in dehydration

19. _____ Stimulates osteoclastic activity

20. _____ Posterior pituitary hormone that is involved in the release of milk from the breast (the milk let-down reflex)

21. _____ Also called *vasopressin*

22. _____ A decrease in plasma calcium is the stimulus for its release

READ THE DIAGRAM

Directions: Referring to the diagram, fill in the spaces with the correct numbers. Some numbers may be used more than once. See text, pp. .

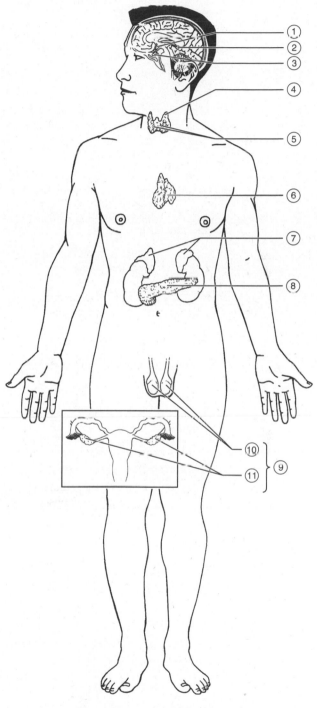

1. _____ Gland that contains the alpha and beta cells of the islets of Langerhans

2. _____ Gonads that secrete estrogen and progesterone

3. _____ A deficiency of this gland causes diabetes insipidus.

4. _____ Secretes T_3, T_4, and calcitonin

5. _____ Secretes releasing hormones

6. _____ Secretes ADH and oxytocin

7. _____ A deficiency of this gland causes cretinism and myxedema.

8. _____ A deficiency of this gland causes hypocalcemic tetany.

9. _____ Called the *hypophysis*

10. _____ The hypothalamus and this gland compose the hypothalamic-hypophyseal portal system.

11. _____ Secretes tropic hormones such as ACTH, TSH, growth hormone, and gonadotropins

12. _____ Secretes the catecholamines, epinephrine, and norepinephrine

13. _____ Secretes the steroids—glucocorticoids, mineralocorticoids, and androgens

14. _____ Target gland for ACTH

15. _____ Target gland for TSH

16. _____ Secretes insulin and glucagon

17. _____ Target gland (female) for FSH and LH

18. _____ Target gland (male) for gonadotropins

19. _____ A low plasma level of calcium stimulates these glands to secrete PTH.

20. _____ A high plasma level of calcium stimulates this gland to secrete calcitonin.

21. _____ Secretion of this gland contributes to the fight-or-flight response.

22. _____ Gland that secretes cortisol, aldosterone, and testosterone

23. _____ Hypersecretion of this gland causes Cushing's syndrome.

24. _____ Hypersecretion of this gland causes Graves' disease.

25. _____ Hypersecretion of this gland is associated with exophthalmos.

26. _____ Gonad that secretes testosterone

27. _____ Gland that uses iodine to synthesize its hormones

28. _____ Gland that plays an important role in immunity and involutes, or shrinks, after puberty

29. _____ Secretes hormones concerned with sugar, salt, and sex

30. _____ A cone-shaped gland located in the brain; called the *biological clock*

31. _____ Gland that secretes the hormone that lowers blood glucose level

32. _____ Gland that secretes FSH and LH

33. _____ Hormone that secretes lactogenic hormone

34. _____ Gland that is the target gland for the releasing hormones

35. _____ Gland that can develop a goiter

36. _____ Gland that controls BMR

37. _____ This gland secretes a hormone that promotes labor.

38. _____ Hypersecretion of this gland can cause gigantism (childhood) or acromegaly (adult).

39. _____ Refers to the male and female gonads

40. _____ Gland that secretes both steroids and catecholamines

COLORING

Directions: Color the appropriate areas on the illustration on the previous page as indicated below.

1. Color the pituitary gland *red.*

2. Color the gland that secretes insulin and glucagon *blue.*

3. Color the gland that secretes T_3 and T_4 *yellow.*

4. Color the target gland of FSH and LH *purple.*

5. Color the suprarenal glands *pink.*

6. Color the parathyroid glands *black.*

7. Color the thymus gland *orange.*

READ THE DIAGRAM

Directions: Referring to the diagram, fill in the spaces with the correct numbers. Some numbers may be used more than once. See text, pp. 271-274.

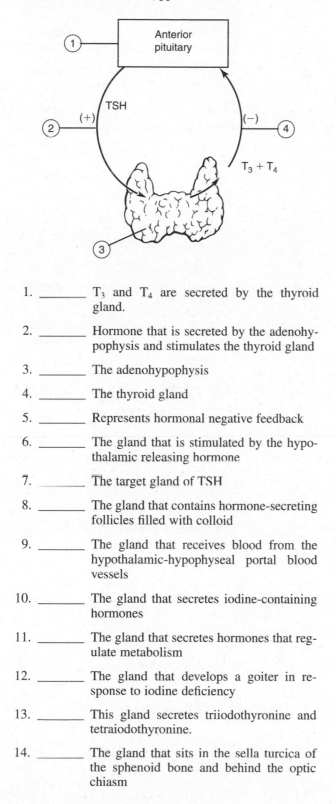

1. _____ T_3 and T_4 are secreted by the thyroid gland.

2. _____ Hormone that is secreted by the adenohypophysis and stimulates the thyroid gland

3. _____ The adenohypophysis

4. _____ The thyroid gland

5. _____ Represents hormonal negative feedback

6. _____ The gland that is stimulated by the hypothalamic releasing hormone

7. _____ The target gland of TSH

8. _____ The gland that contains hormone-secreting follicles filled with colloid

9. _____ The gland that receives blood from the hypothalamic-hypophyseal portal blood vessels

10. _____ The gland that secretes iodine-containing hormones

11. _____ The gland that secretes hormones that regulate metabolism

12. _____ The gland that develops a goiter in response to iodine deficiency

13. _____ This gland secretes triiodothyronine and tetraiodothyronine.

14. _____ The gland that sits in the sella turcica of the sphenoid bone and behind the optic chiasm

15. _____ The gland that is anterior to the trachea, a respiratory structure

16. _____ The gland that can easily be palpated on physical examination

17. _____ Administration of Synthroid, a thyroxine drug, causes a decrease in the secretion of TSH from this gland.

18. _____ Two lobes of this gland are connected by an isthmus.

READ THE DIAGRAM

Directions: Referring to the diagram, fill in the spaces with the correct numbers. Some numbers may be used more than once. See text, pp. 271-274.

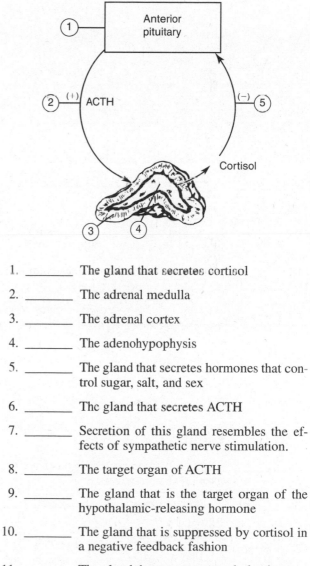

1. _____ The gland that secretes cortisol

2. _____ The adrenal medulla

3. _____ The adrenal cortex

4. _____ The adenohypophysis

5. _____ The gland that secretes hormones that control sugar, salt, and sex

6. _____ The gland that secretes ACTH

7. _____ Secretion of this gland resembles the effects of sympathetic nerve stimulation.

8. _____ The target organ of ACTH

9. _____ The gland that is the target organ of the hypothalamic-releasing hormone

10. _____ The gland that is suppressed by cortisol in a negative feedback fashion

11. _____ The gland that secretes catecholamines

12. _____ The adrenal cortical hormone that suppresses the adenohypophyseal secretion of ACTH

13. _____ The gland that secretes steroids

14. _____ The gland that secretes glucocorticoids, mineralocorticoids, and androgens

15. _____ This gland secretes epinephrine and norepinephrine (NE).

16. _____ Hypersecretion of this gland causes Cushing's syndrome.

17. _____ The administration of prednisone, a steroid similar to a natural glucocorticoid, suppresses the secretion of ACTH from this gland.

18. _____ Hypersecretion of this gland causes pheochromocytoma.

READ THE DIAGRAM

Directions: Referring to the diagram, fill in the spaces with the correct numbers. Some numbers may be used more than once. See text, pp. 274-276.

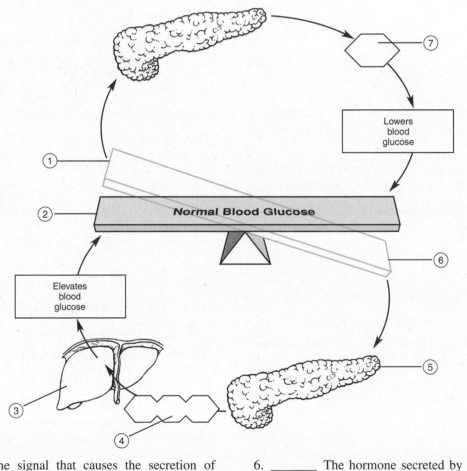

1. _____ The signal that causes the secretion of glucagon

2. _____, _____ The signal that diminishes the secretion of glucagon

3. _____ The signal that causes the secretion of insulin

4. _____, _____ The signal that diminishes the secretion of insulin

5. _____ The part of the board that indicates a declining blood glucose

6. _____ The hormone secreted by the pancreas in response to a decreased blood glucose

7. _____ The part of the board that indicates a rising blood glucose

8. _____ The part of the board that indicates blood glucose about a half-hour after a meal

9. _____ The organ in which glucagon stimulates glycogenolysis, thereby increasing blood glucose

10. _____ The hormone secreted by the pancreas in response to an elevated blood glucose

MATCHING

Endocrine Disorders

Directions: Match the following terms to the most appropriate definition by writing the correct letter in the space provided. Some terms may be used more than once. See text, pp. 264-277.

A. diabetes mellitus
B. Addison's disease
C. Cushing's syndrome
D. goiter
E. exophthalmos
F. Graves' disease
G. diabetes insipidus
H. pheochromocytoma
I. myxedema
J. tetany
K. acromegaly
L. giantism
M. dwarfism
N. cretinism

1. _____ This results from an iodine deficient diet, which causes a hyperplasia or overgrowth of the thyroid gland.

2. _____ Results from a deficiency of insulin; the person becomes hyperglycemic and glucosuric.

3. _____ Condition that is caused by a deficiency of adrenocortical steroids

4. _____ Results from a deficiency of ADH that causes the person to urinate up to 25 L/day of pale urine

5. _____ Condition that is treated with replacement doses of insulin

6. _____ The adult form of hypothyroidism that is treated with thyroid hormone

7. _____ Antithyroid drugs and surgery are used to treat this form of hyperthyroidism.

8. _____ An infant born with a deficiency of thyroid hormones develops this condition, characterized by a severe delay in both physical and mental development.

9. _____ A deficiency of parathyroid hormone causes this life-threatening hypocalcemic condition.

10. _____ Condition that is caused by hypersecretion of growth hormone in an adult (after the fusion of the epiphyseal discs)

11. _____ Condition that is caused by a hypersecretion of somatotropic hormone in a child

12. _____ Condition that is caused by hyposecretion of somatotropic hormone in a child

13. _____ Condition that results from a tumor of the adrenal medulla, which causes a very high (and dangerous) blood pressure

14. _____ Describes the bulging or protruding eyes that characterize hyperthyroidism

15. _____ A person who takes high doses of steroids over an extended time will develop this group of signs and symptoms.

16. _____ If untreated, this deficiency disease will progress to adrenal shock.

17. _____ If untreated, this condition progresses to ketoacidosis.

18. _____ Characterized by sustained muscle contraction and an inability to ventilate (breathe)

19. _____ Requires replacement doses of steroids, especially mineralocorticoid

20. _____, _____ Requires replacement doses of T_3 and T_4 (need 2 lines for the answers)

TELL A STORY

Ms. Chvostek's Face Is A-Twitching

Directions: Use these words to complete the story. See text, pp. 269-271.

(+) Chvostek's sign
PTH
kidneys
tetany
osteoclastic activity
(+) Trousseau's sign
laryngospasm
carpal spasm
parathyroid
intestine

Plasma levels of Ca^{2+} are regulated very closely. As the plasma Ca^{2+} level decreases, the hormone _____ is secreted by the _____ glands. This hormone stimulates the _____ to reabsorb Ca^{2+} from the urine and the _____ to increase the absorption of dietary Ca^{2+}. Most important, the hormone stimulates _____; this action causes the Ca^{2+} to leave the bone and enter the plasma, thereby elevating the plasma Ca^{2+} level. So why is Ms. Chvostek's face a-twitching? Ms. Chvostek had her parathyroid glands surgically removed by accident (during a thyroidectomy). She subsequently developed hypocalcemia; this condition is manifested clinically as _____ (a sustained muscle contraction). Two "muscle" signs of hypocalcemic tetany are _____ and _____. The Ca^{2+} imbalance also increases the nerve irritability, manifested clinically as _____ and _____.

SIMILARS AND DISSIMILARS

Directions: Circle the word in each group that is least similar to the others. Indicate the similarity of the three words on the line below each question.

1. beta cells   islets of   aldosterone   alpha
             Langerhans   cells

   _____

2. myxedema   hyperthyroidism   Cushing's   toxic
                                    syndrome   goiter

   _____

3. ADH   kidney   water balance   adenohypophysis

   _____

4. mineralocorticoid   androgen

      glucocorticoid   epinephrine

   _____

5. cortisol   aldosterone   glucagon   testosterone

   _____

6. ACTH   prolactin   growth hormone   ADH

   _____

7. ADH   oxytocin   epinephrine   vasopressin

   _____

8. PTH T_3 TSH   thyroxine

   _____

9. adenohypophysis   pancreas

      pituitary gland   neurohypophysis

   _____

10. FSH   gonadotropins   insulin   LH

    _____

11. glucagon   adrenal cortex   pancreas   insulin

    _____

12. steroids   catecholamines

       epinephrine   norepinephrine

    _____

13. T_4 insulin   triiodothyronine   thyroxine

    _____

14. prolactin   pancreas   milk   lactogenic hormone

    _____

15. PTH   calcium   salt retention   osteoclastic
                                            activity

    _____

16. diabetes insipidus   glucagon   ADH   polyuria

    _____

17. insulin   pancreas   ADH   hyperglycemia
       deficiency   deficiency

    _____

18. giantism   acromegaly   goiter   iodine

    _____

19. (+) Chvostek's   hypocalcemic   cretinism   carpal
       sign   tetany   spasm

    _____

20. estrogen   testosterone   ACTH   progesterone

    _____

21. cortisol   glucocorticoid   Cushing's   carpal
                                      syndrome   spasm

    _____

22. T_3 thyroxine   ADH   calcitonin

    _____

23. ADH   kidney   water balance   adenohypophysis

    _____

Part II: Putting It All Together

MULTIPLE CHOICE

Directions: Choose the correct answer.

1. Which of the following is least true of or related to an endocrine gland?
   a. It secretes hormones.
   b. They are ductless glands.
   c. Secretions are transported by the blood.
   d. All hormones are steroids.

2. The adenohypophysis
   a. secretes only steroids.
   b. is controlled by the hypothalamus.
   c. is controlled by the posterior pituitary gland.
   d. requires iodine for the synthesis of TSH.

3. Complete this series: CRH → ACTH →
   a. T_3 and T_4
   b. cortisol
   c. insulin
   d. ADH

4. Complete this series: TRH → TSH →
   a. T_3 and T_4
   b. cortisol
   c. insulin
   d. ADH

5. Which of the following is true of insulin?
   a. secreted by the posterior pituitary gland
   b. deficiency causes diabetes insipidus
   c. secreted in response to decreased blood glucose
   d. helps regulate blood glucose

6. Which of the following is not true of glucagon?
   a. lowers blood glucose level
   b. synthesized by the pancreas
   c. synthesized by the islets of Langerhans
   d. opposes the action of insulin

7. Which of the following statements is true about ACTH?
   a. It is a releasing hormone.
   b. It is secreted by the hypothalamus.
   c. It stimulates the adrenal cortex to secrete cortisol.
   d. It is released in response to elevated cortisol levels in the blood.

8. Which of the following is most related to FSH and LH?
   a. They are secreted by the pancreas.
   b. They are neurohypophyseal hormones.
   c. They regulate the blood glucose level.
   d. They are tropic hormones aimed at the ovaries and testes.

9. The ketoacidosis of diabetes mellitus is
   a. caused by hyperglycemia.
   b. a consequence of glucosuria.
   c. caused by the rapid and incomplete breakdown of fatty acids.
   d. caused by an insulin-induced gluconeogenesis.

10. Hyperglycemia, glucosuria, and ketoacidosis are caused by
    a. excess secretion of ACTH by the anterior pituitary gland.
    b. a deficiency of ADH.
    c. a deficiency of insulin.
    d. hypersecretion of catecholamines by the adrenal medulla.

11. What characteristic is shared by both insulin and glucagon?
    a. Both are steroids.
    b. Both raise the blood glucose level.
    c. Both lower the blood glucose level.
    d. Both are secreted by the pancreas.

12. ACTH, TSH, and prolactin
    a. are secreted by the pancreas.
    b. are releasing hormones.
    c. stimulate the adrenal cortex to secrete cortisol.
    d. are synthesized by the adenohypophysis.

13. Which of the following is most related to the adrenal cortex?
    a. It is the target gland of ACTH.
    b. It regulates calcium homeostasis.
    c. It secretes PTH.
    d. It is embedded within the thyroid gland.

14. BMR, iodine, myxedema, and Graves' disease are all
    a. concerned with T_3 and T_4.
    b. concerned with the regulation of the plasma calcium level.
    c. characterized by hypoglycemia.
    d. concerned with adenohypophyseal secretions.

15. Which statement is true about ADH?
    a. It is secreted by the anterior pituitary gland.
    b. It helps to regulate blood volume.
    c. A deficiency causes hypervolemia (expanded blood volume).
    d. A deficiency causes oliguria (a decrease in urine excretion).

16. A deficiency of aldosterone
    a. causes a decrease in blood volume.
    b. causes diabetes insipidus.
    c. causes hyperglycemia, hypernatremia, and hyperkalemia.
    d. is called *Cushing's syndrome*.

Chapter **14** Endocrine System

17. Aldosterone
    a. stimulates the reabsorption of potassium by the kidney.
    b. causes the excretion of sodium in the urine.
    c. excretes water, causing an increase in the output of urine.
    d. stimulates the reabsorption of sodium and water by the kidney.

18. PTH
    a. stimulates osteoclastic activity.
    b. lowers the plasma calcium level.
    c. elevates the blood glucose level.
    d. requires iodine for its synthesis.

19. Thyroxine
    a. is an adenohypophyseal tropic hormone.
    b. helps regulate body metabolism.
    c. elevates plasma calcium through its effect on bones, kidneys, and digestive tract.
    d. stimulates the anterior pituitary gland to secrete TSH.

20. A patient has been receiving a large dose of prednisone, a glucocorticoid for the relief of arthritic pain, for 6 months. What is the most serious concern when he suddenly stops taking the medication?
    a. His arthritic pain will recur.
    b. He will develop an acute adrenal insufficiency.
    c. He will continue to exhibit the symptoms of adrenal cortical excess (Cushing's syndrome).
    d. His adrenal cortex will oversecrete cortisol.

21. Prolactin
    a. determines BMR.
    b. is a gonadotropin.
    c. is also called lactogenic hormone.
    d. is the major hormonal component of the milk letdown reflex.

22. The pancreas
    a. secretes two hormones, one that lowers blood glucose and another that raises blood glucose.
    b. is under the direct control of the adenohypophysis.
    c. secretes both ADH and oxytocin.
    d. secretes all hormones that affect blood glucose.

23. The beta cells of the islets of Langerhans
    a. secrete a hormone that elevates blood glucose.
    b. are insulin-secreting pancreatic cells.
    c. use iodide salts in the synthesis of pancreatic hormones.
    d. secrete steroids.

24. Which of the following is most apt to induce virilizing effects?
    a. androgenic hormones
    b. iodide salts
    c. hormones with osteoclastic activity
    d. neurohypophyseal hormones

25. Which of the following words includes this information: extremities and enlargement?
    a. polydipsia
    b. polyphagia
    c. acromegaly
    d. euthyroid

26. Which of the following words includes this information: in the blood, glucose, and excess?
    a. glucosuria
    b. polyuria
    c. hyperglycemia
    d. ketoacidosis

27. Which of the following includes this information: in the blood, Ca^{2+} and below?
    a. hypoglycemia
    b. hypocalcemia
    c. calciuria
    d. acromegaly

28. Which of the following words includes this information: poisoning, condition of, and T_3/T_4?
    a. hypothyroidism
    b. hyperparathyroidism
    c. thyrotoxicosis
    d. adenohypophysis

29. Which of the following describes the adenohypophysis?
    a. neural
    b. exocrine
    c. hypofunctional
    d. glandular

30. Which of the following statements is true regarding the aging process and the endocrine system?
    a. With aging, most glands decrease their hormonal secretion, resulting in severe clinical disorders, such as diabetes mellitus and Addison's disease.
    b. Pheochromocytoma is the result of the normal aging process.
    c. Although most glands decrease their hormonal secretions, normal aging does not cause deficiency states.
    d. Most older adults suffer from osteoporosis and tetany.

CASE STUDY

J.C., a 17-year-old high school senior, is a star football player and has just been awarded an athletic scholarship to college. About 8 months ago, he began using anabolic steroids to improve his athletic performance.

1. The steroids most resemble the secretions of which gland?
   a. pancreas
   b. adrenal medulla
   c. posterior pituitary gland
   d. adrenal cortex

2. What steroid-induced effect was J.C. expecting?
   a. increased muscle mass (bulking up) and strength
   b. decreased appetite
   c. decreased red blood cell production
   d. increased sex drive

PUZZLE

Hint: 'Roid Rage

Directions: Perform the following functions on the Sequence of Words that follows. When all the functions have been performed, you are left with a word or words related to the hint. Record your answer in the space provided.

Functions: Remove the following:

1. Three classifications of adrenal steroids

2. The salt-retaining mineralocorticoid

3. Two adrenal medullary catecholamines

4. A group of symptoms that are caused by excess secretion of the adrenal cortex

5. The condition caused by a chronic hypofunction of the adrenal cortex

6. The cation that is eliminated in the urine in response to aldosterone

7. The cation that is reabsorbed from the urine in response to aldosterone

8. The rounded face and supraclavicular fat pad caused by excess cortisol

9. Disorder caused by excess secretion of the adrenal medulla (catecholamines)

10. ACTH stimulates the adrenal cortical secretion of this hormone.

11. Blood glucose-regulating hormones (two) secreted by the beta cells of the islets of Langerhans

12. Gland that secretes hormones that regulate BMR

13. Gland that is the primary regulator of plasma Ca^{2+}

14. Gland that secretes releasing hormones

15. Target gland of releasing hormones

16. Gland that secretes ADH and oxytocin

17. Classification of FSH and LH

Sequence of Words

S O D I U M M O O N F A C E A D E N O H Y P O P
H Y S I S M I N E R A L O C O R T I C O I D C U S H
I N G S S Y N D R O M E N E U R O H Y P O P H Y S
I S P H E O C H R O M O C Y T O M A T H Y R O I D
C O R T I S O L G O N A D O T R O P I N S E P I N E
P H R I N E P A R A T H Y R O I D G L U C A G O N
D R U G - I N D U C E D A G G R E S S I O N G L U C
O C O R T I C O I D B U F F A L O H U M P N O R E
P I N E P H R I N E A D D I S O N S D I S E A S E A
L D O S T E R O N E A N D R O G E N H Y P O T H
A L A M U S K⁺I N S U L I N

Answer: _____

Blood

OBJECTIVES

1. Describe three functions of blood.
2. Describe the composition of blood, including:
   - The three types of blood cells: erythrocytes, leukocytes, and thrombocytes.
   - The formation of blood cells.
3. Explain the composition, characteristics, and functions of red and white blood cells and platelets, including the breakdown of red blood cells and the formation of bilirubin.
4. Identify the steps of hemostasis.
5. Describe the four blood types.
6. Describe the Rh factor.

Part I: Mastering the Basics

MATCHING

Blood and Blood Tests

Directions: Match the following terms to the most appropriate definition by writing the correct letter in the space provided. Some terms may be used more than once. See text, pp. 284-294.

A. plasma	J. intrinsic factor
B. hemopoiesis	K. erythropoietin (EPO)
C. hematocrit (HCT)	L. hematology
D. hemoglobin	M. red bone marrow
E. thrombocytes	N. blast cell
F. leukocytes	O. erythropoiesis
G. bone marrow biopsy	P. thrombopoiesis
H. erythrocytes	Q. leukopoiesis
I. differential count	

1. _____ Process of platelet production by the bone marrow

2. _____ Blood test that indicates the percentage of blood cells in a sample of blood; assumed to be the percentage of red blood cells (RBCs)

3. _____ Component of the RBC to which the oxygen is attached

4. _____ Protein secreted by the stomach that is necessary for the absorption of vitamin B_{12}

5. _____ Process of white blood cell (WBC) production

6. _____ Process of blood cell production

7. _____ Blood test that indicates the percentage of each type of WBC in a sample of blood

8. _____ Process of RBC production by the bone marrow

9. _____ Hormone that stimulates the production of RBCs

10. _____ Liquid portion of the blood

11. _____ A sample of developing blood cells is withdrawn from the sternum or iliac crest; this procedure can detect abnormal blood cells.

12. _____ Part of the blood that contains the gamma globulins, fibrinogen, and albumin

13. _____ An immature cell

14. _____ Part of the bone that makes most of the blood cells

15. _____ Tissue that is impaired in myelosuppression

16. _____ The study of blood

17. _____ Platelets

18. _____ WBCs

19. _____ RBCs

Hematocrit

Directions: Label the two parts of the blood on the diagram. See text, pp. 285-286.

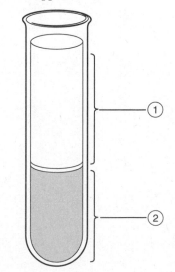

In the spaces provided, indicate whether the following describes plasma (P) or the formed elements (FE).

1. _____ Contains the erythrocytes, leukocytes, and thrombocytes

2. _____ Contains the plasma proteins

3. _____ Contains the albumin, clotting factors, and globulins

4. _____ Removal of the clotting factors from this part of the blood produces serum

5. _____ A decrease in this part is indicative of anemia.

6. _____ Contains the RBCs, WBCs, and platelets

7. _____ An increase in this part of the blood lowers the hematocrit.

8. _____ Normally colored pale yellow

9. _____ Normally colored red

10. _____ This part of the blood decreases in the dehydrated state.

Hematocrit Measurements

Draw an arrow on the test tubes (A, B, C) indicating the normal hematocrit. Then draw arrows indicating the hematocrits for the following cases. Record your explanation for the change in hematocrit.

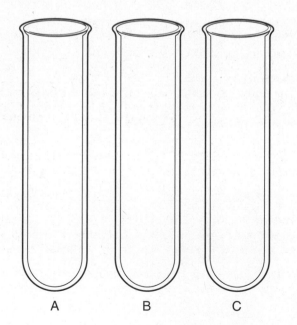

A   B   C

A. A patient has been vomiting for 1 week (stomach "bug") and is severely dehydrated.

B. An underweight 20-year-old vegan has been dieting for 3 months, losing 20 pounds. She was diagnosed with iron deficiency anemia.

C. A hospitalized patient was overhydrated with IV fluid and developed a "dilutional" anemia.

_____

_____

_____

READ THE DIAGRAM

Directions: Referring to Figure 15-2 in the textbook, fill in the blanks with the number of the column described by the statements below. See text, p. 287.

1. _____ The maturation of the granulocytes

2. _____ The maturation of the oxygen-carrying blood cells

3. _____ The maturation of the cells that play a role in hemostasis

4. _____, _____, and _____ the maturation of the leukocytes

5. _____ The maturation of a myeloblast into a highly phagocytic cell

6. _____ The fragmentation of the megakaryocyte into the thrombocyte

7. _____ Depression of the cells in this column causes granulocytopenia and infection.

8. _____ Depression of the cells in this column causes thrombocytopenia and bleeding.

9. _____ The reticulocyte is an immature cell.

10. _____ Depression of this column causes anemia.

MATCHING

Red Blood Cells, White Blood Cells, and Platelets

Directions: Indicate whether the following are most related to the red blood cell (RBC), white blood cell (WBC), or platelet (P) by writing the correct letter(s) in the space provided. See text, pp. 287-296.

1. _____ Small, highly phagocytic granulocyte

2. _____ Antigens A and B

3. _____ Delivers oxygen to the cells in the body

4. _____ A deficiency causes petechiae formation and bleeding.

5. _____ Life span is about 120 days

6. _____ Breaks down into protein, bilirubin, and iron

7. _____ Includes lymphocytes and monocytes

8. _____ Includes neutrophils, basophils, and eosinophils

9. _____ Is filled primarily with hemoglobin

10. _____ The immature cell is the reticulocyte.

11. _____ Derived from the megakaryocyte

12. _____ Is concerned primarily with phagocytosis

13. _____ Is composed of granulocytes and agranulocytes

14. _____ Leukocytosis and leukopenia

15. _____ Its rapid breakdown causes hyperbilirubinemia.

16. _____ Is decreased in anemia

17. _____ Primarily concerned with hemostasis

18. _____ Synthesized in response to EPO

19. _____ Thrombocytopenia and hemorrhage

20. _____ Granulocytopenia and infection

21. _____ Contributes to the formation of pus

22. _____ The Rh factor

23. _____ Its rapid breakdown causes jaundice.

24. _____ Shift to the left

25. _____ Monitored through changes in the HCT.

26. _____ Stickiness and plug

27. _____ Involved in a hemolytic blood transfusion reaction

28. _____ Intrinsic and extrinsic factors are necessary for its synthesis.

29. _____ Requires iron for its synthesis and function

30. _____ Segs, polys, (PMNs), band cells

TELL A STORY

Polly Cythemia and Her Oxygen

Directions: Complete the story by using the words below. Some words may be used more than once. See text, p. 290.

polycythemia	bone marrow
erythropoietin	kidneys
hypoxemic	red blood cells
oxygen	myelosuppression
anemia	

An older adult patient with emphysema was chronically hypoxemic and developed an elevated hematocrit. Why, you ask? In response to the low tissue levels of oxygen,

the _____ secrete the hormone _____.

This hormone travels via the blood to the _____,

where it stimulates the formation of _____. The increased amount of hemoglobin delivers more

_____ to the tissues. Chronic hypoxemia, therefore, causes a secondary _____ and an increase in hematocrit. There are other clinical conditions that involve erythropoietin. Cancer patients often develop

a chemotherapy-induced _____ and anemia.

Patients in kidney failure develop _____ (a deficiency of RBCs) because of a deficiency of erythropoietin. Interestingly, athletes who train "at altitude" become

mildly _____ and therefore secrete erythropoietin; in response, they, too, develop a secondary

_____.

Breakdown of the Red Blood Cell

Directions: Referring to the illustration (Figure 15-6 in the textbook), indicate the breakdown products of the red blood cell by writing the correct numbers in the blanks. See text, pp. 290, 291.

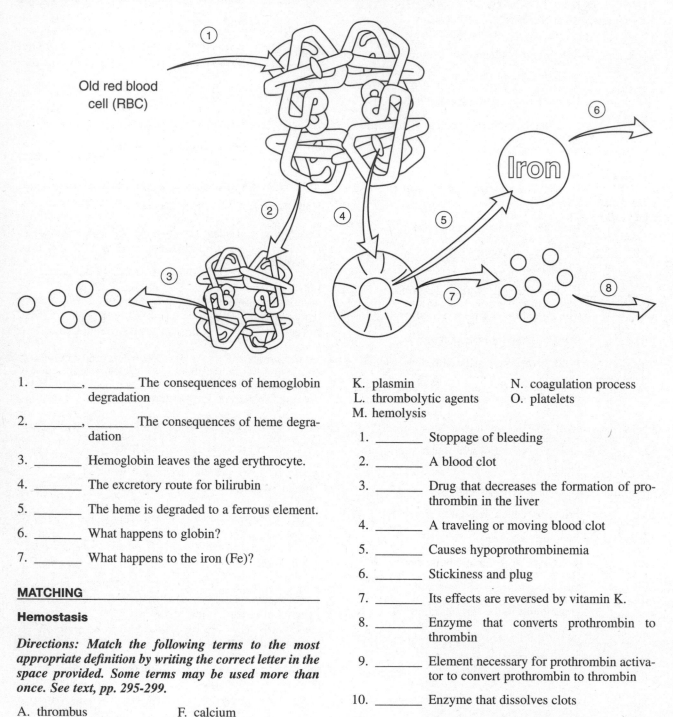

Old red blood cell (RBC)

Iron

1. _____, _____ The consequences of hemoglobin degradation

2. _____, _____ The consequences of heme degradation

3. _____ Hemoglobin leaves the aged erythrocyte.

4. _____ The excretory route for bilirubin

5. _____ The heme is degraded to a ferrous element.

6. _____ What happens to globin?

7. _____ What happens to the iron (Fe)?

MATCHING

Hemostasis

Directions: Match the following terms to the most appropriate definition by writing the correct letter in the space provided. Some terms may be used more than once. See text, pp. 295-299.

A. thrombus
B. prothrombin activator
C. fibrin threads
D. embolus
E. hemostasis

F. calcium
G. thrombin
H. heparin
I. Coumadin
J. fibrinolysis

K. plasmin
L. thrombolytic agents
M. hemolysis

N. coagulation process
O. platelets

1. _____ Stoppage of bleeding

2. _____ A blood clot

3. _____ Drug that decreases the formation of prothrombin in the liver

4. _____ A traveling or moving blood clot

5. _____ Causes hypoprothrombinemia

6. _____ Stickiness and plug

7. _____ Its effects are reversed by vitamin K.

8. _____ Enzyme that converts prothrombin to thrombin

9. _____ Element necessary for prothrombin activator to convert prothrombin to thrombin

10. _____ Enzyme that dissolves clots

11. _____ Drug that prolongs the prothrombin time (PT)

12. _____ Drugs that are called *clot busters*

13. _____ Protein strands that form the blood clot

14. _____ Derived from the megakaryocyte

15. _____ An anticoagulant that removes thrombin from the clotting process

16. _____ Bursting of red blood cells

17. _____ Enzyme that activates fibrinogen to fibrin

18. _____ Thrombocytes

19. _____ Refers to the series of reactions that results in the formation of a blood clot

20. _____ The target of aspirin

21. _____ Process that dissolves a blood clot

Blood Coagulation

Directions: In the space provided, indicate the number that is described.

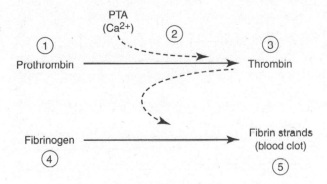

1. _____ Activation of prothrombin

2. _____ Catalyst that activates prothrombin

3. _____ The enzyme that converts fibrinogen to fibrin

4. _____ Plasmin works here.

5. _____ The vitamin K–dependent clotting factor that is synthesized by the liver

6. _____ The clotting factor that is affected by Coumadin

7. _____ The clotting factor that is the primary target of heparin

8. _____ Fibrinolytic drugs work here.

9. _____ A decrease in this clotting factor causes a prolonged PT.

10. _____ Clot retraction occurs here.

Directions: Referring to the previous diagram, answer these questions.

1. Describe how Coumadin acts as an anticoagulant.

   _____

2. Describe how heparin works as an anticoagulant.

   _____

3. Citrate (found in donor blood) removes calcium. Explain why excessive amounts of donor citrated blood cause bleeding in the recipient.

   _____

4. Explain why the administration of excessive fibrinolytic drugs may cause the patient to hemorrhage.

   _____

TELL A STORY

Got a Clot

Directions: Complete the story by using the words below. See text, pp. 295-299.

prothrombin	hypoprothrombinemia
Coumadin	heparin
prothrombin time	thrombin
antithrombin	

Mr. Flea Bitis was admitted to the hospital with a diagnosis of deep vein thrombosis (DVT), a life-threatening condition associated with clot formation in the veins of

the legs. He was given the drug _____ by intravenous (IV) infusion. This anticoagulant is an

_____ agent. The drug-induced removal of

_____ prevents the conversion of fibrinogen to fibrin threads (the clot). On the following day, he was

also given _____, an oral anticoagulant. This anticoagulant works by preventing the hepatic utilization

of vitamin K in the synthesis of _____. The goal of oral anticoagulant therapy is to cause

_____; this is confirmed by a prolonged

_____, a laboratory test used to monitor therapy. The intravenous anticoagulant was discontinued within the first week; however, the oral anticoagulant was continued for several months until the thrombotic condition was resolved.

MATCHING

Blood Types

Directions: Match the following terms to the most appropriate definition by writing the correct letter in the space provided. Some terms may be used more than once. See text, pp. 298-301.

A. type A+
B. type B+
C. type AB+
D. type O–

E. Rh factor
F. erythroblastosis fetalis
G. hemolysis
H. kernicterus

1. _____ The universal recipient

2. _____ The RBCs of this blood type contain neither the A antigen nor the B antigen.

3. _____ The plasma of this blood type contains both anti-A antibodies and anti-B antibodies.

4. _____ The plasma of this blood type contains neither anti-A antibodies nor anti-B antibodies.

5. _____ A person with this blood type can receive (by transfusion) type A, type B, type AB, or type O blood.

6. _____ The universal donor

7. _____ A person with this blood type can receive (by transfusion) only type O– blood.

8. _____ The administration of mismatched blood causes this serious condition.

9. _____ The positive and negative signs (e.g., A+, A–) refer to this antigen.

10. _____ Severe hemolytic reaction in the fetus that is caused by an antigen-antibody reaction involving the blood cells of the mother and fetus

11. _____ A serious neurological complication of erythroblastosis fetalis that results in severe developmental delay

12. _____ Blood type that contains only anti-B antibodies

13. _____ Blood type that contains only anti-A antibodies

14. _____ A person with type A– blood can receive this type of blood.

15. _____ A person with type B– blood can receive this type of blood.

16. _____ Blood type that includes the A antigen, the B antigen, and the Rh factor

17. _____ Consequence of administering type A+ blood to a patient who is type AB–

TELL A STORY

The Yellow Baby

Directions: Complete the story by using the words below. See text, pp. 299-302.

erythroblastosis fetalis
hemolysis
Rh antigens

hyperbilirubinemia
anti-RH antibodies

Ms. Billie Rubin (blood type A–) gave birth to a baby girl (type A+), a second daughter. Within 4 hours after birth, the baby appeared jaundiced; the jaundice intensified during the next 12 hours. What happened? Ms. Rubin has

type A– blood and _____, having been sensitized from the first pregnancy. During the second pregnancy, the maternal antibodies slipped across the placenta into the baby and attacked the _____ of the baby's RBCs, thereby causing agglutination and

_____. The ruptured RBCs release bilirubin, causing _____ and jaundice. This condition is called _____.

MATCHING

Red Blood Cell Review with the Anemias

Directions: Match the following terms to the most appropriate definition by writing the correct letter in the space provided. Some terms may be used more than once. It may be helpful to refer to the Medical Terminology and Disorders table in the text (p. 303). See text, pp. 287-291.

A. hemolytic anemia
B. folic acid deficiency anemia
C. pernicious anemia
D. sickle cell anemia

E. iron deficiency anemia
F. aplastic anemia
G. anemia of chronic renal failure

1. _____ Caused by impaired function of the parietal cells in the stomach; they are unable to secrete intrinsic factor.

2. _____ Reticulocytes are usually absent.

3. _____ A hereditary anemia that causes the red blood cells to form a rigid crescent shape

4. _____ Endemic in a low-income population

5. _____ A diabetic with end-stage renal disease (diabetic nephropathy)

6. _____ Anemia that characterizes erythroblastosis fetalis

7. _____ A megaloblastic anemia that is treated with vitamin B_{12} injections

8. _____ Treated with ferrous sulfate

9. _____ An infant who drinks only whole milk for the first year of life is likely to develop this type of anemia.

10. _____ Accompanied by granulocytopenia and thrombocytopenia

11. _____ The most painful of the anemias

12. _____ Myelosuppression

13. _____ Anemia associated with blood loss and occult blood in stools (positive guaiac test)

14. _____ Anemia that is characterized by jaundice

15. _____ The lack of intrinsic factor impairs the absorption of extrinsic factor.

16. _____ A megaloblastic anemia that is commonly seen in pregnant women and patients with alcoholism

17. _____ The anemia most often associated with kernicterus

18. _____ A hypochromic, microcytic anemia that is often caused by a chronic slow-bleeding lesion

19. _____ Often seen in a cancer patient who is being treated with powerful cytotoxic drugs and radiation

20. _____ Following a severe hemorrhage, a person will experience this type of anemia.

SIMILARS AND DISSIMILARS

Directions: Circle the word in each group that is least similar to the others. Indicate the similarity of the three words on the line below each question.

1. erythrocyte   platelet   thrombus   leukocyte

_____

2. neutrophil   basophil   eosinophil   megakaryocyte

_____

3. megakaryocyte   WBC   platelet   thrombocyte

_____

4. neutropenia   leukocytosis
   thrombocytopenia   anemia

_____

5. hemolysis   bilirubin   EPO   jaundice

_____

6. pernicious   iron deficiency   hemostasis   sickle cell

_____

7. EPO   aspirin   Coumadin   heparin

_____

8. granulocyte   hemoglobin   bilirubin Fe^{2+}

_____

9. erythrocyte   "poly"   RBC   reticulocyte

_____

10. prothrombin   fibrinogen   factor VIII   bilirubin

_____

11. leukocytosis   kernicterus
    hemolysis   hyperbilirubinemia

_____

12. granulocyte   icterus   phagocytosis   neutrophil

_____

13. neutrophils   polymorpho-   reticulocyte   "polys"
                  nuclear
                  leukocytes
                  (PMNs)

_____

14. Rh   A   B   AB

_____

15. hemolysis   vitamin K   fibrinogen   prothrombin

16. prolonged PT   anticoagulation

    jaundice   hypoprothrombinemia

_____

17. leukocytosis   infection   anemia   shift to the left

_____

18. vitamin B_{12} gastric   pernicious   kernicterus
             mucosa   anemia

_____

19. myelosuppression   thrombocytopenia

    infection   bleeding

_____

20. sickle cell   aplastic   hemolytic   hemostasis

_____

21. EPO   erythropoiesis   aplastic O_2 concentration

_____

22. jaundice   fibrin strands   blood clot   thrombus

_____

23. EPO   blood clot   embolus   thrombus

_____

24. PTA Ca^{2+} Fe^{2+} thrombin

_____

25. neutropenia   granulocytopenia

    thrombocytopenia   leukopenia

_____

26. icterus   jaundice   hyperbilirubinemia   cyanosis

_____

27. Rh(−)   coagulation   platelet plug   vascular
                                     spasm

_____

28. fibrinolysis   tissue   plasmin   hemopoiesis
              plasminogen
              activator
              (tPA)

_____

29. epistaxis   bleeding   petechiae   thrombosis

_____

30. vitamin B_{12} intrinsic   extrinsic   hemophilia
                  factor   factor

_____

31. heparin   antithrombin   vitamin K   anticoagulant

_____

Part II: Putting It All Together

MULTIPLE CHOICE

Directions: Choose the best answer.

1. Polymorphs (polys), segs, and band cells are
   a. platelets that become sticky and form a plug.
   b. erythrocytes that carry oxygen throughout the body.
   c. neutrophils.
   d. antibody-secreting agranulocytes.

2. Which of the following is a true statement?
   a. Several key blood clotting factors are synthesized in the liver.
   b. EPO is secreted by the bone marrow.
   c. Prothrombin is synthesized by the kidney.
   d. The spleen utilizes vitamin K to synthesize prothrombin.

3. Myelosuppression
   a. diminishes the numbers of blood cells.
   b. causes a deficiency of clotting factors.
   c. causes hypoprothrombinemia and a prolonged PT.
   d. activates plasminogen.

4. Vasospasm, platelet plug, and blood coagulation are most related to which process?
   a. agglutination
   b. phagocytosis
   c. fibrinolysis
   d. hemostasis

5. Carbon monoxide binds to
   a. granulocytes, causing infection.
   b. hemoglobin, causing hypoxemia.
   c. platelets, causing hypoxemia.
   d. bone marrow, causing bleeding.

6. Which of the following descriptions is most related to an erythrocyte?
   a. a blood cell that participates in hemostasis
   b. a blood cell that becomes sticky on activation
   c. a hemoglobin-containing cell that carries oxygen
   d. a phagocytic cell

7. Which of the following is the stimulus for the release of EPO?
   a. elevated serum bilirubin level
   b. lowered levels of oxygen in the blood
   c. increase in the amount of iron in the blood
   d. decrease in the synthesis of intrinsic factor

8. Which of the following is least descriptive of bilirubin?
   a. originates in hemoglobin
   b. liberated from heme
   c. stored in the liver and used in the synthesis of fibrinogen
   d. excreted in the bile

9. Erythroblastosis fetalis is most likely to occur in which of the following situations?
   a. mother is type A–; baby is type A+
   b. mother is type B+; baby is type B–
   c. mother is type B–; baby is type B–
   d. mother is type AB+; baby is type B+

10. Which combination is correct?
    a. hemolysis and leukopenia
    b. hypoxemia and jaundice
    c. hypercoagulability and bleeding
    d. hypoprothrombinemia and bleeding

11. Granulocytopenia is most related to
    a. leukocytosis and fever.
    b. platelet deficiency and bleeding.
    c. neutropenia and infection.
    d. anemia and cyanosis.

12. Which of the following is not a function of plasma proteins?
    a. carry oxygen
    b. maintain blood volume
    c. transport hormones and substances such as bilirubin and drugs
    d. fight infection

13. Rapid hemolysis causes
    a. hypoxemia and flushing of the face.
    b. myelosuppression and bleeding.
    c. hyperbilirubinemia and jaundice.
    d. leukopenia and infection.

14. A person in chronic kidney failure is anemic because
    a. he cannot secrete adequate intrinsic factor.
    b. he cannot tolerate iron-rich foods.
    c. his kidneys do not secrete adequate EPO.
    d. his diseased kidneys excrete folic acid.

15. Tissue plasminogen activator (tPA) is a drug that activates plasmin and, therefore,
    a. suppresses bone marrow activity.
    b. prevents the hepatic synthesis of prothrombin.
    c. blocks the hepatic use of vitamin K in the synthesis of prothrombin.
    d. dissolves clots.

16. Which of the following patients is most likely to benefit from an injection of vitamin K?
    a. patient with pernicious anemia
    b. patient who has iron deficiency anemia
    c. the granulocytopenic patient
    d. the hypoprothrombinemic patient

17. Albumin, globulins, and fibrinogen are all
    a. clotting factors.
    b. plasma proteins.
    c. anticoagulants.
    d. blood types.

18. Which condition is caused by venous stasis?
    a. bleeding
    b. thrombosis
    c. jaundice
    d. hemophilia

19. Which of the following patients is most likely to have a low reticulocyte count?
    a. the patient who is hypoprothrombinemic
    b. the patient who is myelosuppressed
    c. the patient who is hypoxic and cyanotic
    d. the patient who is hyperbilirubinemic and jaundiced

20. Which of the following is true of all these words: neutropenia, thrombocytopenia, and granulocytopenia?
    a. states of deficiency
    b. leukocyte function
    c. types of anemias
    d. most common response to infection

21. Which of the following is descriptive of both thrombus and embolus?
    a. petechiae
    b. ecchymoses
    c. blood clot
    d. icterus

22. This word literally means "lover of blood" and refers to the bleeding tendency of the person with
    a. icterus
    b. myelosuppression
    c. hemophilia
    d. leukemia

23. Which of the following is a correct combination?
    a. hyperbilirubinemia, icterus, and EPO
    b. Coumadin, hypoprothrombinemia, anticoagulation
    c. EPO, thrombocytopenia, thrombosis
    d. heparin, antiplatelet, fibrinolysis

24. Expanded blood volume and/or diminished erythropoiesis is most apt to
    a. decrease hematocrit.
    b. increase susceptibility to infection.
    c. cause thrombosis in the deep veins of the legs.
    d. cause jaundice in the hemoglobin molecule.

25. Which of the following best describes the role of iron?
    a. binds reversibly to oxygen
    b. holds the globin chains together
    c. prevents hemolysis
    d. transports CO_2

26. The recipient of a blood transfusion has type A– blood. Which of the following is a compatible donor blood type?
    a. A+
    b. B–
    c. O–
    d. AB–

27. Which of the following is true of all these words: hemostasis, hematology, hemolytic, and hematuria?
    a. All refer to diagnostic studies concerning blood.
    b. All refer to blood.
    c. All refer to blood dyscrasias (disorders).
    d. All refer to consequences of anemia.

CASE STUDY

B.R., a 72-year-old widowed farmer, lives alone on a fixed income. He went to his physician complaining of being tired all the time. His diet consists primarily of tea and toast. A blood test revealed the following: a hemoglobin level of 9 g/dL and a hematocrit value of 27. He had a WBC count of 9000/mm^3 and a normal WBC differential count. His platelet count was 300,000/mm^3. No other signs and symptoms were noted.

1. B.R.'s medical history and blood studies are most suggestive of which condition?
    a. hemophilia
    b. infection
    c. anemia
    d. platelet deficiency

2. What is the most likely cause of his medical condition?
    a. iron-poor diet
    b. drug-induced hemolysis
    c. thrombocytopenia
    d. infection

3. Which blood studies support the diagnosis?
    a. abnormal WBC count
    b. abnormal platelet count
    c. abnormally low hemoglobin and HCT
    d. elevated HCT

4. What is the cause of the fatigue (feeling tired all the time)?
    a. excessive phagocytic activity
    b. overstimulation of the bone marrow
    c. reduced oxygenation of the tissues
    d. bleeding

5. Which of the following blood studies ruled out an underlying infection?
    a. HCT
    b. hemoglobin
    c. normal WBC and WBC differential counts
    d. platelet count

PUZZLE

Hint: "Why the Yellow Fellow?"

Directions: Perform the following functions on the Sequence of Words that follows. When all the functions have been performed, you are left with a word or words related to the hint. Record your answer in the space provided.

Functions: Remove the following:

1. Another name for the RBC

2. Refers to bone marrow depression

3. Three granulocytes

4. The anticoagulant that acts as an antithrombin agent

5. The organ that synthesizes EPO

6. A vitamin K–dependent clotting factor

7. A deficiency of platelets that causes bleeding

8. The breakdown of RBCs

9. The protein that fills the RBC and is concerned with the transport of oxygen

10. The ratio of the formed elements of the blood to the total blood volume

11. The organ that synthesizes prothrombin

12. A, B, AB, and O are located on the outer membrane surface of the RBC

13. Another term for blood clotting

14. An enzyme that causes fibrinolysis

15. Means "little yellow bird," a reference to jaundice

16. Refers to the production of blood cells; includes erythropoiesis, leukopoiesis, and thrombopoiesis

17. Fragments of a megakaryocyte (two names)

18. Immature RBC

Sequence of Words

LIVERICTERUSHEMATOCRITTHRO
MBOCYTESPROTHROMBINBASOPH
ILHYPERBILIRUBINEMIAPLASMIN
THROMBOCYTOPENIARETICULOCY
TEMYELOSUPPRESSIONANTIGENER
YTHROCYTEHEMOLYSISCAUSESHE
MOGLOBINHEMOPOIESISNEUTROP
HILKIDNEYPLATELETSJAUNDICEEO
SINOPHILHEPARINCOAGULATION

Answer: _____

16 Anatomy of the Heart

OBJECTIVES

1. Describe the location of the heart.
2. Name the three layers and the covering of the heart.
3. Explain the function of the heart as two separate pumps.
4. Identify the four chambers and great vessels of the heart.
5. Explain the functions of the four heart valves.
6. Describe the physiological basis of the heart sounds.
7. Describe blood flow through the heart.
8. List the vessels that supply blood to the heart.
9. Identify the major components of the heart's conduction system.

Part I: Mastering the Basics

MATCHING

Location, Layers, and Chambers of the Heart

Directions: Match the following terms to the most appropriate definition by writing the correct letter in the space provided. Some terms may be used more than once. See text, pp. 308-313.

A. endocardium
B. pericardium
C. atria
D. ventricles
E. epicardium
F. angina pectoris
G. right atrium
H. coronary arteries
I. left atrium
J. right ventricle
K. myocardial infarction (MI)
L. coronary veins
M. myocardium
N. base
O. left ventricle
P. great vessels
Q. precordium
R. apex
S. pericardial cavity
T. cardiology

1. _____ Sling-like structure that supports the heart

2. _____ Delivers oxygenated blood to the myocardium

3. _____ Smooth, shiny, innermost lining of the heart

4. _____ Chamber that receives unoxygenated blood from the venae cavae

5. _____ Chamber that pumps unoxygenated blood to the lungs through the pulmonary artery

6. _____ Term that includes the venae cavae, pulmonary artery, pulmonary veins, and aorta

7. _____ Primary pumping chambers of the heart

8. _____ Chamber that receives oxygenated blood from the lungs through four pulmonary veins

9. _____ Outermost layer of the heart

10. _____ Chamber that pumps oxygenated blood into the systemic circulation

11. _____ The myocardium is the thickest in this chamber.

12. _____ Death of the heart muscle caused by occlusion of a coronary artery

13. _____ Receiving chambers of the heart

14. _____ Drains unoxygenated blood from the myocardium

15. _____ Chest pain usually caused by impaired flow of blood through the coronary arteries

16. _____ Area on the anterior chest that overlies the heart and great vessels

17. _____ The left anterior descending (LAD) is a branch of these.

18. _____ Hardest working chamber

19. _____ Layer of the heart that contains the contractile proteins actin and myosin

20. _____ The epicardium is part of this structure.

21. _____ Upper flat portion of the heart that is located at the level of the second rib

22. _____ Collection of fluid or blood in this space that causes an external compression of the heart (called *cardiac tamponade*)

23. _____ Study of the heart

24. _____ Pointed end of the heart that is located at the level of the fifth intercostal space

Blood Flow Through the Heart

Directions: Referring to the figure, write the correct number in the spaces below. Some words may be used more than once. See text, pp. 311-316.

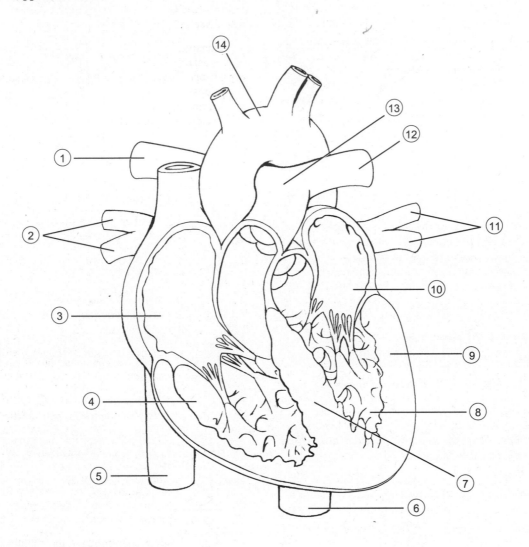

1. _____ The superior and inferior venae cavae empty blood into this structure.

2. _____ The left ventricle pumps blood into this structure.

3. _____ The right ventricle pumps blood into this structure.

4. _____ The pulmonary veins empty blood into this structure.

5. _____ This separates the two pumping chambers.

6. _____, _____ The bicuspid valve separates these two structures.

7. _____, _____ The mitral valve separates these two structures.

8. _____ The aortic valve separates the aorta from this structure.

9. _____ The pulmonic valve separates the right ventricle from this structure.

10. _____, _____ The tricuspid valve separates these two structures.

11. _____ Blood flows through the tricuspid valve into this chamber.

12. _____ Blood is pumped through the pulmonic valve into this structure.

13. _____ Blood flows through the mitral valve into this chamber.

14. _____ Blood is pumped through the right semilunar valve into this structure.

15. _____ Blood flows through the right atrioventricular (AV) valve from this chamber.

16. _____ Blood flows through the left AV valve into this structure.

17. _____ An incompetent mitral valve allows the retrograde flow of blood from the left ventricle into this structure.

18. _____ A stenotic (narrowed) pulmonic valve causes hypertrophy of the muscle of this chamber.

19. _____ An incompetent right semilunar valve allows the retrograde flow of blood from this structure into the right ventricle.

20. _____ A stenotic left semilunar valve causes hypertrophy of the myocardium of this chamber.

21. _____, _____ The main pulmonary artery bifurcates into these two arteries.

22. _____, _____ Oxygenated blood returns to the left atrium by these four veins.

23. _____ The inferior vena cava

24. _____ An extension of #14; it is the descending aorta

25. _____ The hardest working muscular structure in the heart

COLORING

Directions: Color or mark the appropriate areas on the illustration on the previous page as indicated below.

1. Color all the structures that carry oxygenated blood *red.*

2. Color all the structures that carry unoxygenated blood *blue.*

3. Color the ventricular myocardium *yellow.*

4. Put an *X* over each AV valve.

5. Place a *Y* over each semilunar valve.

MATCHING

Heart Valves and Sounds

Directions: Match the following terms to the most appropriate definition by writing the correct letter in the space provided. Some terms may be used more than once. See text, pp. 313-316.

A. tricuspid valve G. left ventricle
B. bicuspid valve H. incompetent valve
C. gallop rhythm I. chordae tendineae
D. stenosis J. S_1
E. pulmonic valve K. S_2
F. aortic valve

1. _____ Semilunar valve through which blood leaves the right ventricle

2. _____ Atrioventricular valve on the right side of the heart

3. _____ Called the *mitral valve*

4. _____ Leaky valve that allows backflow of blood

5. _____ The first heart sound (lubb)

6. _____ Exit valve that sees only oxygenated blood

7. _____ Tough bands that attach the AV valves to the ventricular walls

8. _____ Refers to the narrowing of a valve

9. _____ Extra heart sounds (S_3, S_4) that sound like a racing horse

10. _____ Valve that prevents the backflow of blood from the aorta into the left ventricle

11. _____ Heart sound created by the closure of the AV valves at the beginning of ventricular contraction

12. _____ Entrance valve that sees only oxygenated blood

13. _____ Valve that prevents the backflow of blood from the left ventricle into the left atrium

14. _____ AV valve between the left atrium and the left ventricle

15. _____ Exit valve that sees only unoxygenated blood

16. _____ Valve that prevents the backflow of blood from the pulmonary artery

17. _____ Heart sound created by the closure of the semilunar valves at the beginning of ventricular relaxation

18. _____ Semilunar valve through which blood leaves the left ventricle

19. _____ Entrance valve that sees only unoxygenated blood

127

COMPLETE THE TABLE

Blood Flow

Directions: Complete the table by filling in the names of the chambers and great vessels as blood flows through the heart. Example: Blood flows from the right atrium → right ventricle → pulmonary artery. See text, pp. 311-316.

From →	To →	To
right atrium	right ventricle	pulmonary artery
pulmonary capillaries		
		aorta
	left atrium	
venae cavae		
left atrium		
	pulmonary capillaries	
		pulmonary arteries

MATCHING

Conduction System

Directions: Match the terms to the most appropriate definition by writing the correct letter in the space provided. Some terms may be used more than once. See text, pp. 318-321.

A. AV node
B. sinoatrial (SA) node
C. Purkinje fibers
D. QRS complex
E. tachycardia
F. normal sinus rhythm (NSR)
G. bundle of His
H. ectopic focus
I. ventricular fibrillation
J. electrocardiogram (ECG)
K. P wave
L. P-R interval
M. bradycardia
N. T wave

1. _____ Life-threatening dysrhythmia that causes the ventricular myocardium to quiver in an uncoordinated and ineffective way

2. _____ Pacemaker of the heart

3. _____ Heart rate below 60 beats/min

4. _____ Specialized conduction tissue in the interventricular septum; divides into the right and left branches

5. _____ Place where the electrical signal normally arises

6. _____ The electrical signal spreads from the conduction tissue in the atria to this structure.

7. _____ Area outside the SA node that gives rise to an electrical signal

8. _____ These fast-conducting fibers spread the electrical signal throughout the ventricular wall.

9. _____ Record of the electrical activity of the heart (a heart chart)

10. _____ Heart rate higher than 120 beats/min

11. _____ ECG deflection that indicates atrial depolarization

12. _____ ECG deflection that indicates ventricular depolarization

13. _____ ECG deflection that represents ventricular repolarization

14. _____ ECG recording that represents the time it takes for the electrical signal to travel from the SA node to the ventricles

15. _____ Electrical event that stimulates ventricular myocardial contraction

16. _____ Electrical event that stimulates atrial contraction

17. _____ Lengthening of this time indicates heart block.

18. _____ Electrical activity appears normal, and the signal arises within the SA node.

Conduction System

Directions: Referring to the figure, enter the numbers in the blanks. Some words may be used more than once. See text, pp. 318-321.

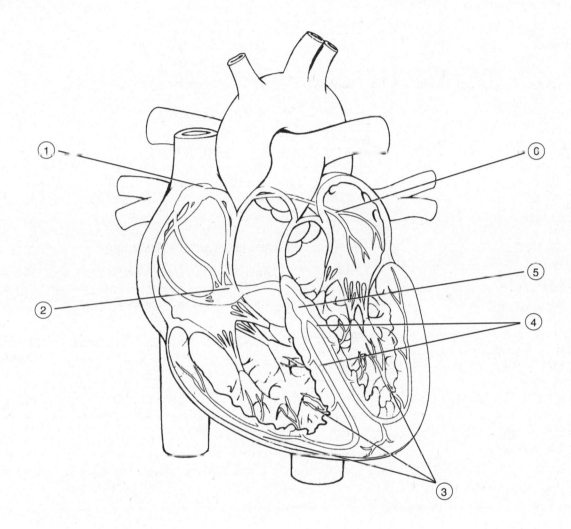

1. _____ Pacemaker of the heart

2. _____ Specialized conduction tissue in the inter-ventricular septum; divides into the right and left branches

3. _____ Where the electrical signal arises in normal sinus rhythm (NSR)

4. _____ A bundle

5. _____ The electrical signal spreads from the conduction tissue in the atria to this structure.

6. _____ Fast-conducting fibers that spread the electrical signal throughout the ventricular wall

7. _____ Where the electrical signal arises in nodal rhythm

8. _____ Left atrial conduction fibers

ORDERING

Blood Flow Through the Heart

Directions: Indicate blood flow through the heart by writing the correct term in the space provided. Two of the steps are given. See text, pp. 311-316.

bicuspid valve (mitral) aorta
right ventricle tricuspid valve
pulmonary trunk left atrium
pulmonary veins (four) pulmonic semilunar valve
left ventricle aortic semilunar valve
pulmonary arteries
(right and left)

1. Right atrium

2.

3.

4.

5.

6.

7. Pulmonary capillaries

8.

9.

10.

11.

12.

13.

RED AND BLUE

Directions: Color the box red if it carries oxygenated blood. Color the box blue if it carries unoxygenated blood.

1. ❑ pulmonary artery

2. ❑ right atrium

3. ❑ mitral valve

4. ❑ tricuspid valve

5. ❑ left ventricle

6. ❑ left atrium

7. ❑ venae cavae

8. ❑ aorta

9. ❑ right semilunar valve

10. ❑ aortic valve

11. ❑ pulmonic valve

12. ❑ right ventricle

13. ❑ pulmonary veins

SIMILARS AND DISSIMILARS

Directions: Circle the word in each group that is least similar to the others. Indicate the similarity of the three words on the line below each question.

1. mitral   SA node   tricuspid   pulmonic semilunar

_____

2. AV node   His-Purkinje   precordium   pacemaker

_____

3. atrium   ventricle   pumping chamber   pericardium

_____

4. epi-   endo-   vaso-   myo-

_____

5. aorta   venae cavae   right atrium   pulmonary veins

_____

6. tricuspid   AV valve   mitral   precordium

_____

7. P wave S_1, S_2 QRS complex   P-R interval

_____

8. left anterior   troponin   left   circumflex
   descending   coronary
                                  artery

_____

9. actin   troponin   myosin   creatine
                                     phosphokinase
                                     (CPK)

_____

10. CPK   Lactate   pericardial   serum
             dehydrogenase   fluid   troponin
             (LDH)

_____

11. lubb-dupp S_1 fibrillation S_2

_____

12. angina S_1, S_2 cardiac   myocardial
                                tamponade   infarction

_____

13. ventral   thoracic   cranial   mediastinum
    cavity   cavity   cavity

_____

14 cardiac   ECG   pacemaker   AV valve
   impulse

_____

15. NSR   SA node   80 beats/min   nodal
                                         rhythm

_____

16. QRS   ventricular   His-Purkinje   pericardium
    complex   depolarization

_____

17. ventricular   left-to-right   interventricular   SA
    septal   shunt   septum   node
    defect

_____

18. AV valve   great   chordae   valve
                 vessels   tendineae   cusps

_____

19. apex   precordium   pulse   base

_____

Part II: Putting It All Together

MULTIPLE CHOICE

Directions: Choose the correct answer.

1. Which of the following is not related to the location of the heart?
   a. thoracic cavity
   b. mediastinum
   c. pleural cavity
   d. precordium

2. Oxygenated blood is found in which structure?
   a. right ventricle
   b. right atrium
   c. pulmonary artery
   d. pulmonary veins

3. What is the cause of the heart sounds "lubb-dupp"?
   a. vibrations caused by closure of valves
   b. firing of the SA node
   c. movement of blood through the valves
   d. flow of blood through the coronary vessels

4. Which of the following is an example of tachycardia?
   a. NSR
   b. heart rate of 85 beats/min
   c. heart rate of 40 beats/min
   d. rapid rate (120 beats/min)

5. Which of the following statements is true?
   a. The left ventricular myocardium pumps blood to the pulmonic circulation.
   b. The left ventricular myocardium is thicker than the right ventricular myocardium because it works harder.
   c. The right ventricular myocardium pumps blood to the systemic circulation.
   d. The right ventricular myocardium pumps blood into the aorta.

6. Blood flow through the coronary arteries is greatest
   a. when the semilunar valves are open.
   b. during ventricular contraction.
   c. when the AV valves are closed.
   d. during ventricular relaxation.

7. Levels of cardiac enzymes (LDH, CPK, aspartate, serum transaminase) and troponin increase
   a. whenever a person experiences angina.
   b. when myocardial cells are damaged and leak their contents.
   c. during ventricular contraction when coronary blood flow diminishes.
   d. in the presence of valvular dysfunction.

131

8. The AV node
   a. slows the electrical signal as it moves from the atrium to the ventricles.
   b. is the pacemaker of the heart.
   c. is located in the upper right atrium.
   d. sends the electrical signal to the SA node.

9. The precordium is
   a. located within the mediastinum.
   b. located within the pericardial space.
   c. the anterior chest overlying the heart and great vessels.
   d. the space within the chambers of the heart.

10. The mitral valve is
    a. a semilunar valve that sees oxygenated blood.
    b. the right AV valve.
    c. a cuspid valve that sees unoxygenated blood.
    d. a cuspid valve.

11. Which of the following statements is least descriptive of the ventricular myocardium?
    a. contraction and relaxation
    b. pumping chambers
    c. QRS complex
    d. location of the SA node

12. Which of the following separates the right and left ventricles?
    a. pericardium
    b. interventricular septum
    c. chordae tendineae
    d. papillary muscles

13. Both the tricuspid and mitral valves
    a. "see" unoxygenated blood.
    b. are located in the left heart.
    c. are semilunar valves.
    d. have chordae tendineae attached to their cusps.

14. Which of the following describes the LAD artery and the circumflex artery?
    a. parts of the venae cavae
    b. parts of the pulmonic circulation
    c. branches of the left coronary artery
    d. branches of the descending aorta

15. The Purkinje fibers
    a. send electrical signals to the bundle of His.
    b. are the specialized cells of the SA node.
    c. are rapid conduction fibers that supply the ventricular myocardium.
    d. are responsible for the P wave (ECG).

16. The P-R interval (ECG)
    a. is shortened in heart block.
    b. represents atrial depolarization.
    c. represents the time it takes the electrical signal to travel from the atrium to the ventricles.
    d. represents ventricular depolarization.

17. Ventricular contraction is least effective during
    a. a heart rate of 70 beats/min.
    b. NSR.
    c. ventricular fibrillation.
    d. exercise.

18. Which of the following is responsible for the opening and closing of heart valves?
    a. oxygen saturation of the blood in the heart chambers
    b. hematocrit (HCT)
    c. pressure within the heart chambers
    d. tugging on the cusps

19. Blood flows from the venae cavae to the pulmonary artery. Which structure is not included in this blood path?
    a. right ventricle
    b. tricuspid valve
    c. pulmonic semilunar valve
    d. pulmonary veins

20. Blood flows from the right ventricle to the left semilunar valve. Which structure is not in this blood path?
    a. pulmonary veins
    b. right atrium
    c. bicuspid valve
    d. left atrium

21. Which of the following is a direct consequence of the QRS complex?
    a. contraction of the chordae tendineae
    b. second heart sound (dupp)
    c. myocardial contraction
    d. myocardial relaxation

22. Describe the valvular response to the relaxation of the right ventricle.
    a. The tricuspid valve opens.
    b. The pulmonic valve opens.
    c. The right AV valve snaps closed.
    d. The semilunar valve in the right heart opens.

23. Which of the following is a correct valvular position when the atria are contracting?
    a. The tricuspid valve is closed.
    b. The pulmonic valve is open.
    c. Both semilunar valves are open.
    d. Both AV valves are open.

24. Which of the following is true of S_2?
   a. a heart sound known as lubb
   b. indicative of gallop rhythm
   c. occurs during the beginning of ventricular contraction with the closure of the AV valves
   d. heard at the beginning of ventricular relaxation

25. Which of the following describes the heart as a double pump?
   a. atria and ventricles
   b. right and left
   c. AV and semilunar
   d. endocardium and myocardium

26. What is the origin of the electrical signal (cardiac impulse) in a patient in normal sinus rhythm (NSR)?
   a. SA node
   b. AV node
   c. an atrial ectopic site
   d. His-Purkinje system

27. Bradycardia and tachycardia refer to abnormal
   a. amounts of oxygen in the blood flowing through the cardiac chambers.
   b. changes in the contractile force of the myocardium.
   c. heart rates.
   d. ECG readings, specifically the shape of the QRS complex.

28. Contraction of the papillary muscles
   a. closes the AV valves.
   b. opens the AV valves.
   c. opens the bicuspid and tricuspid valves.
   d. exerts tension on the chordae tendineae.

29. Spontaneous depolarization is most descriptive of
   a. actin and myosin interaction.
   b. a pacemaker cell.
   c. contraction of the papillary muscles.
   d. valvular function.

30. Which of the following words includes this information: disease, muscle, and heart?
   a. electrocardiogram
   b. pericarditis
   c. cardiomyopathy
   d. tachycardia

CASE STUDY

M.S. went to bed about 11 PM after a busy evening of entertaining friends and family. He was awakened at 2 AM with chest pain that radiated to his left shoulder, arm, and fingers. His son took him to the emergency department, where he was immediately given O_2 by mask and nitroglycerin, a vasodilator drug. His chest pain was relieved in about 30 minutes. An ECG revealed evidence of myocardial ischemia but no evidence of a myocardial infarction. He was admitted to the cardiac intensive care unit for further evaluation.

1. Which of the following is the most likely cause of the pain?
   a. an insufficient amount of oxygen available to the myocardium
   b. constriction of the heart by a swollen pericardium
   c. an inflamed valve
   d. strain of the shoulder and arm muscles

2. Which of the following terms refers to chest pain?
   a. heart attack
   b. myocardial infarction
   c. angina pectoris
   d. myocardial necrosis

3. Which of the following explains why nitroglycerin helps relieve anginal pain?
   a. It deadens the sensory nerves in the heart.
   b. It decreases the amount of "pumping" work by the heart.
   c. It causes the heart muscle to tetanize.
   d. It dissolves blood clots.

PUZZLE

Hint: Achy Breaky Heart

Directions: Perform the following functions on the Sequence of Words that follows. When all the functions have been performed, you are left with a word or words related to the hint. Record your answer in the space provided.

Functions: Remove the following:

1. Two semilunar heart valves

2. Three AV (cuspid) heart valves

3. Three layers of the heart

4. Heart's sling

5. Two pumping chambers

6. Parts of the conduction system (three)

7. The left ventricle ejects blood into this large artery.

8. The right ventricle ejects blood into this artery.

9. Blood vessels that nourish the myocardium

10. Chamber that receives oxygenated blood from the pulmonary veins

11. Two myocardial contractile proteins

12. Heart sounds

13. Area of the anterior chest wall overlying the heart and great vessels

14. Three cardiac enzymes

Sequence of Words

PRECORDIUMLEFTATRIUMASTLU
BBDUPPEPICARDIUMSANODEPU
LMONARYTRUNKAVNODEACTINE
NDOCARDIUMLDHAORTICMITRA
LLEFTVENTRICLECPKMYOCARDI
UMACUTEMYOCARDIALINFARCT
IONCORONARYARTERIESPULMO
NICMYOSINRIGHTVENTRICLEAO
RTAHIS-PURKINJESYSTEMANGIN
APECTORISTRICUSPIDPERICARD
IUMBICUSPID

Answers: _____ and _____

BODY TOON

Hint: A Musical Performance

Answer: organ recital

17 Function of the Heart

OBJECTIVES

1. Define *cardiac cycle* with respect to systole and diastole.
2. Describe the autonomic innervation of the heart, including sympathetic and parasympathetic innervation.
3. Define cardiac output, including the following:
   - Describe the effect of Starling's law of the heart on cardiac output.
   - Describe the inotropic effect on cardiac output.
   - Explain how changes in heart rate and/or stroke volume change cardiac output.
4. Define specific clinical vocabulary used to describe cardiac function, including:
   - Define preload (end diastolic volume), and explain how it affects cardiac output.
   - Define afterload, and identify the major factor that determines afterload.
   - Define chronotropic, inotropic, and dromotropic effects.
5. Define heart failure, and differentiate between right-sided and left-sided heart failure.

Part I: Mastering the Basics

MATCHING

Myocardial Function

Directions: Match the terms to the most appropriate definition by writing the correct letter in the space provided. Some terms may be used more than once. See text, pp. 325-330.

A. cardiac output
B. systole
C. diastole
D. stroke volume
E. cardiac cycle
F. positive inotropic effect
G. Starling's law of the heart
H. cardiac reserve
I. heart rate

1. _____ Increase in the strength of myocardial contraction that occurs without stretching of the heart

2. _____ Increase in the strength of myocardial contraction that occurs when the heart is stretched

3. _____ Phase of the cardiac cycle that refers to contraction of the heart muscle

4. _____ Sequence of events that occurs in the heart in one beat

5. _____ Beats/min (bpm)

6. _____ Coronary blood flow is greatest during this phase of the cardiac cycle.

7. _____ Determined by heart rate × stroke volume

8. _____ Amount of blood pumped by the heart in 1 minute

9. _____ 70 mL/beat

10. _____ The capacity to increase cardiac output above the resting cardiac output

11. _____ What the ventricles are "doing" when the atrioventricular (AV) valves are closed and the semilunar valves are open

12. _____ Amount of blood pumped by the ventricle per beat

13. _____ Its duration is 0.8 second with a normal resting heart rate.

14. _____ The phase of the cardiac cycle that refers to relaxation of the ventricles

15. _____ Mechanism that allows the myocardium to match venous return and cardiac output on a beat-to-beat basis

16. _____ Phase of the cardiac cycle that shortens most in response to tachycardia

17. _____ What the ventricles are "doing" when the AV valves are open and the semilunar valves are closed

18. _____ 5000 mL/min

MATCHING

Autonomic Innervation of the Heart

Directions: Indicate if the following effects are caused by sympathetic activity (S) or parasympathetic activity (P). See text, pp. 326-328.

1. _____ Increases heart rate and stroke volume

2. _____ Is associated with the fight-or-flight response

3. _____ Increases cardiac output

4. _____ Also called *vagal activity*

5. _____ Causes a positive (+) chronotropic and (+) inotropic effect

6. _____ Excess activity causes bradycardia and heart block

7. _____ Causes a negative (−) chronotropic and (−) dromotropic effect

8. _____ Causes a racing and pounding heart

9. _____ Causes bradydysrhythmias

10. _____ Causes tachydysrhythmias

11. _____ Produces effects similar to epinephrine (adrenaline) and dopamine

12. _____ Effects similar to a vagomimetic drug

13. _____ The resting heart rate is dominated by this branch of the autonomic nervous system.

14. _____ Increases force of myocardial contraction

15. _____ Digoxin slows the heart rate by stimulating this branch of the autonomic nervous system.

MATCHING

Electrical or Muscle Contraction

Directions: Indicate if the following are most related to the electrical signal or cardiac muscle contraction. Fill in the blank with electrical (E) or contraction (C). See text, pp. 326-331.

1. _____ Actin and myosin

2. _____ Inotropic effect

3. _____ Depolarization

4. _____ QRS complex

5. _____ Stroke volume

6. _____ T wave

7. _____ Repolarization

8. _____ Systole

9. _____ Sliding filaments

10. _____ Starling's law of the heart

11. _____ Pacemaker

12. _____ Diastole

13. _____ Dysrhythmia

14. _____ Bundle of His

15. _____ Sarcomere

16. _____ Ectopic focus

17. _____ Electrocardiogram (ECG)

18. _____ Purkinje fibers

MATCHING

Heart Talk… "lubb dupp, lubb dupp"

Directions: Match the terms to the most appropriate definition by writing the correct letter in the space provided. Some terms may be used more than once. See text, pp. 329-331.

A. afterload
B. inotropic effect
C. chronotropic effect
D. end-diastolic volume (EDV)
E. ejection fraction
F. dromotropic effect

1. _____ Forms the basis of Starling's law of the heart

2. _____ The percentage of the EDV that is pumped

3. _____ The resistance or opposition to the flow of blood

4. _____ The amount of blood in the ventricles at the end of its resting phase

5. _____ Change in the speed of the cardiac impulse (electrical signal) that travels through the conduction system of the heart

6. _____ Example: hypertension

7. _____ Venous return

8. _____ Also called *preload*

9. _____ Change in myocardial contractile force that is not caused by stretch

10. _____ Change in heart rate

Directions: In the spaces provided, indicate if the statement describes preload (P) or afterload (A). It should be noted that conditions that affect afterload indirectly affect preload. In this exercise, indicate the initial effect.

1. _____ The volume of blood in the ventricle at the end of diastole

2. _____ The degree of stretch of the ventricular muscle

3. _____ Determined by the amount of resistance offered to the pumped blood

4. _____ An example is aortic valve stenosis.

5. _____ An example is the vasoconstrictor response to phenylephrine, an alpha₁-adrenergic agonist.

6. _____ An effect of dehydration and blood loss

7. _____ EDV

8. _____ An effect of a change in venous return

9. _____ A direct effect of an increase in systemic vascular resistance (SVR)

10. _____ A direct effect of chronic systemic or pulmonary artery hypertension

11. _____ Basis of Starling's law of the heart

12. _____ An effect of an increased central venous pressure (CVP)

13. _____ An effect of coarctation (narrowing) of the aorta

14. _____ A direct effect of pulmonary valve stenosis

MATCHING

Autonomic Receptors of the Heart

Directions: Indicate if the following effects are characteristics of the sympathetic nervous system (S) or the parasympathetic nervous system (P). See text, pp. 331-332.

1. _____ Norepinephrine is the postganglionic neurotransmitter.

2. _____ Activation of muscarinic receptors

3. _____ Activation of the beta$_1$ receptors

4. _____ Vagomimetic effects

5. _____ Acetylcholine (ACh) is the postganglionic neurotransmitter.

6. _____ Activation of cholinergic receptors

MATCHING

Heart Failure

Directions: Indicate if the description is more characteristic of left-sided heart failure (L) or right-sided heart failure (R). See text, pp. 333-335.

1. _____ Three-pillow dyspnea

2. _____ Hepatomegaly (enlarged liver), distended jugular veins, pedal edema

3. _____ Pulmonary edema

4. _____ Consequence of chronic lung disease such as emphysema and asthma

5. _____ Backup within the pulmonary capillaries causing water to accumulate in the lungs

6. _____ Cyanosis, dyspnea, orthopnea

7. _____ Most likely to say "I can't breathe."

8. _____ Most likely to develop in response to chronic systemic hypertension

9. _____ Cor pulmonale

SIMILARS AND DISSIMILARS

Directions: Circle the word in each group that is least similar to the others. Indicate the similarity of the three words on the line below each question.

1. systole   diastole   ejection fraction   sinoatrial
                                                (SA) node

_____

2. ejection   dysrhythmia   bradycardia   tachycardia
   fraction

_____

3. ACh   norepinephrine   vagus   parasympathetic

_____

4. EDV   afterload   preload   Starling's law
                                  of the heart

_____

5. afterload   SVR   tricuspid   vasoconstriction

_____

6. (+) chronotropic effect   tachycardia

   beta$_1$-adrenergic activation   vagomimetic

_____

7. vagomimetic   bradycardia

   muscarinic activation   tachydysrhythmias

_____

8. pulmonary edema   elevated CVP

   dehydration   heart failure

_____

9. vagolytic   antimuscarinic

   increased heart rate   (+) inotropism

_____

10. norepinephrine   beta$_1$-adrenergic
                             activation

    (+) inotropism   bradycardia

_____

11. ejection fraction 30%   (–) inotropism

    heart failure   normal sinus rhythm

_____

12. jugular   pedal   right-sided   poor
    vein   edema   heart   skin
    distention   failure   turgor

_____

13. dyspnea   increased brain   left-sided   angina
               natriuretic   heart failure
               peptide

_____

14. bradycardia   (+) chronotropic effect

    increased heart rate   sympathomimetic

_____

15. myocardial   stroke   60–80 mL/beat   SVR
    contraction   volume

_____

16. cardiac   stroke   % O_2 heart rate
    output   volume   saturation

_____

17. precordial   inotropic   chronotropic   dromotropic

_____

Part II: Putting It All Together

MULTIPLE CHOICE

_____

Directions: Choose the correct answer.

1. Actin and myosin and striated and involuntary are descriptive terms for the
   a. valves.
   b. myocardium.
   c. pericardium.
   d. great vessels.

2. 70 mL/beat × 72 beats/min is the amount of blood that determines the
   a. ejection fraction.
   b. stroke volume.
   c. cardiac output.
   d. cardiac reserve.

3. Which of the following terms describes myocardial contraction and relaxation?
   a. depolarization and repolarization
   b. P wave and T wave
   c. systole and diastole
   d. tachycardia and bradycardia

4. Which of the following occurs during ventricular systole?
   a. Blood is pumped out of the ventricles.
   b. The ventricles fill with blood.
   c. The AV valves open.
   d. The semilunar valves close.

5. The heart drug digitalis stimulates the parasympathetic nerve that supplies the heart. What cardiac effect is expected?
   a. The valves open faster.
   b. The heart rate increases.
   c. The pulse increases.
   d. The heart rate slows.

6. Which of the following increases stroke volume?
   a. (+) chronotropic effect
   b. (+) dromotropic effect
   c. decreased EDV
   d. (+) inotropic effect

7. Stretching of the heart causes the force of myocardial contraction to increase. This stretch effect is called
   a. (+) inotropic effect.
   b. (+) chronotropic effect.
   c. Starling's law of the heart.
   d. (+) dromotropic effect

8. A chronic elevation in afterload, as in systemic hypertension, is most likely to cause
   a. pulmonic valve stenosis.
   b. pedal edema.
   c. left ventricular hypertrophy.
   d. jugular vein distention.

9. If heart rate increases to 170 beats/min,
   a. the duration of diastole decreases.
   b. coronary blood flow increases.
   c. EDV increases.
   d. preload increases.

10. An increase in venous return of blood to the heart
    a. decreases EDV.
    b. increases preload.
    c. decreases cardiac output.
    d. decreases stroke volume.

11. Preload is most related to
    a. heart rate.
    b. Starling's law of the heart.
    c. an inotropic effect.
    d. SA node activity.

12. A drug, such as atropine, that blocks the muscarinic receptors is most likely to
    a. induce a severe bradycardia.
    b. cause a (–) dromotropic effect, leading to heart block.
    c. increase heart rate.
    d. decrease cardiac output.

13. A beta$_1$-adrenergic agonist
    a. causes bradycardia.
    b. increases cardiac output.
    c. decreases stroke volume.
    d. decreases ejection fraction.

14. Norepinephrine
    a. activates muscarinic receptors.
    b. is antagonized by muscarinic blockade.
    c. is a beta$_1$-adrenergic agonist.
    d. blocks beta$_1$-adrenergic receptors.

15. Blockade of the effects of ACh on the heart
    a. decreases cardiac output.
    b. increases heart rate.
    c. causes a prolongation of the P-R interval (heart block).
    d. causes bradydysrhythmias.

16. What happens when the actin and myosin in the ventricles form cross-bridges?
    a. The valves open.
    b. The myocardium contracts.
    c. The heart enters the period of diastole.
    d. The ventricles fill with blood.

17. For the ventricles to fill, the
    a. AV valves must be closed.
    b. semilunar valves must be open.
    c. ventricles must be in diastole.
    d. chordae tendineae must be fully relaxed.

18. If the QRS complex does not develop,
    a. the heart develops a sustained muscle contraction (tetanizes).
    b. the ventricular myocardium does not contract.
    c. all heart valves close.
    d. the ventricular myocardium remains in systole and cannot relax.

19. Which of the following is most likely to increase the ejection fraction?
    a. vagal discharge
    b. activation of the muscarinic receptors
    c. blockade of the beta$_1$-adrenergic receptors
    d. stimulation of the sympathetic nervous system

20. Which of the following is responsible for the stroke volume?
    a. ventricular contraction
    b. P wave
    c. contraction of the chordae tendineae
    d. firing of the SA node

21. Which of the following is an age-related cardiac change?
    a. The resting heart rate increases to 90 beats/min.
    b. Starling's law of the heart is lost.
    c. The sarcoplasmic reticulum loses its ability to store calcium.
    d. The heart muscle cannot respond as vigorously to the demands of exercise.

CASE STUDY

Mr. I. is a 61-year-old man who was admitted to the emergency department because of a sudden and severe increase in blood pressure (205/100 mm Hg). He was complaining of chest pain, a headache in the back of his head, and a sense of not feeling well. Mr. I. has a history of coronary artery disease and evidence of left ventricular hypertrophy on ECG. He admitted to poor drug compliance (not taking his blood pressure medication) for the past 2 weeks. He was subsequently treated with a calcium channel blocker drug that quickly reduced his blood pressure to 155/95 mm Hg. He was discharged and advised to lose weight, exercise, and resume his antihypertensive drug schedule.

1. Both the chest pain and the left ventricular hypertrophy are caused by
   a. fluid accumulation in the pulmonary capillaries and alveoli.
   b. the increased afterload.
   c. pulmonary artery hypertension.
   d. accumulation of fluid in the pericardial space.

2. Which of the following drug effects decreases afterload?
   a. increased venous return
   b. reduction of blood pressure
   c. (+) inotropic effect
   d. (+) dromotropic effect

3. A calcium channel blocker slows heart rate and decreases force of myocardial contraction. Which of the following best describes these effects?
   a. vasopressor and diuresis
   b. (−) chronotropic effect and (−) inotropic effect
   c. (+) inotropic effect, decreased ejection fraction
   d. (−) dromotropic effect and diuresis

PUZZLE

Hint: Ejection Fraction of 30%

Directions: Perform the following functions on the Sequence of Words below. When all the functions have been performed, you are left with a word or words related to the hint. Record your answer below.

Functions: Remove the following:

1. Contractile proteins (two) found in the myocardium

2. Phases of the cardiac cycle that refer to myocardial contraction and relaxation

3. Electrical terms (two) that form the basis of the P wave, QRS complex, and T wave

4. Parasympathetic nerve that supplies the SA and AV nodes

5. Transmitters (two) of the autonomic nerves of the heart

6. Determinants (two) of cardiac output

7. Mechanisms (two) to change force of myocardial contraction

8. Term that refers to a heart rate of >100 beats/min; term that refers to a heart rate of <60 beats/min

9. Terms (two) that refer to the amount of blood in the ventricles at the end of the resting phase

10. Cholinergic and adrenergic receptors on the SA and AV nodes

Sequence of Words

I N O T R O P I C E F F E C T E N D D I A S T O L I C
V O L U M E B E T$_1$ B R A D Y C A R D I A N O R E P
I N E P H R I N E S T R O K E V O L U M E M Y O S
I N H E A R T F A I L U R E V A G U S D E P O L A R
I Z A T I O N A C T I N D I A S T O L E S Y S T O L E
R E P O L A R I Z A T I O N A C E T Y L C H O L I N
E H E A R T R A T E S T A R L I N G S L A W O F T H
E H E A R T T A C H Y C A R D I A M U S C A R I N
I C P R E L O A D

Answer: _____

18 Anatomy of the Blood Vessels

Answer Key: Textbook page references are provided as a guide for answering these questions. A complete answer key was provided for your instructor.

OBJECTIVES

1. Describe the pulmonary and systemic circulations.
2. Describe the structure and function of arteries, capillaries, and veins, including:
   - List the three layers of tissue found in arteries and veins.
   - Explain the functions of conductance, resistance, exchange, and capacitance vessels.
3. List the major arteries of the systemic circulation that are branches of the ascending aorta, aortic arch, and descending aorta.
4. List the major veins of the systemic circulation.
5. Describe the following special circulations: blood supply to the head and brain, hepatic circulation, and fetal circulation.
6. Explain pulse and its use as an assessment tool.

Part I: Mastering the Basics

MATCHING

Arrangement of Blood Vessels

Directions: Match the following terms to the most appropriate definition by writing the correct letter in the space provided. Some terms may be used more than once. See text, pp. 342-344.

A. systemic circulation
B. veins
C. pulmonary circulation
D. venules
E. arteries
F. capillaries
G. arterioles

1. _____ Large blood vessels that carry blood away from the heart

2. _____ Capacitance vessels

3. _____ The path of the blood from the right ventricle of the heart to the lungs and back to the left atrium

4. _____ These tiny blood vessels are composed of a single layer of epithelium and therefore function as exchange vessels.

5. _____ Blood vessels that carry blood back to the heart

6. _____ The path of the blood from the left ventricle of the heart throughout the body and back to the right atrium

7. _____ Small arteries composed primarily of smooth muscle

8. _____ Vessels that connect the arterioles with the venules

9. _____ Small veins that drain the capillaries and converge to form large veins

10. _____ Large vessels that contain valves

11. _____ Resistance vessels

12. _____ The most numerous of the blood vessels

Arteries

Directions: Referring to the figure, write the correct number in the space provided. Some numbers may be used more than once. See text, pp. 346-349.

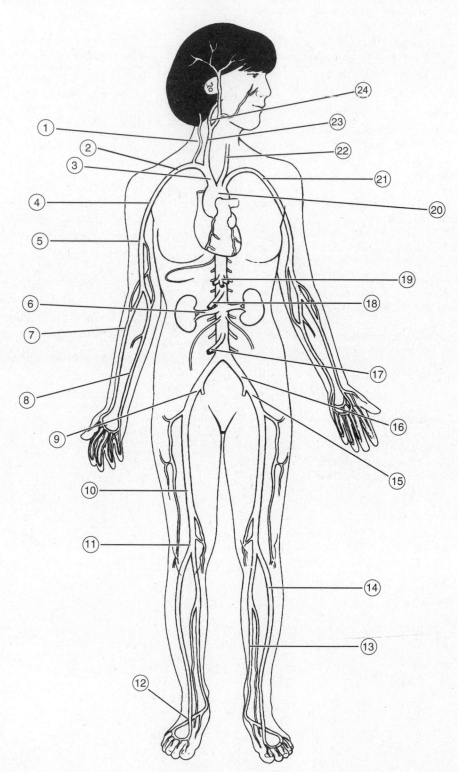

1. _____ Artery that ascends on the posterolateral surface of the neck; both the left and the right arteries merge to form the basilar artery and supply the circle of Willis

2. _____ Artery that has its origin in the right brachiocephalic artery and supplies the circle of Willis

3. _____ Artery that is formed when the descending aorta bifurcates to supply the thigh

4. _____ Artery that is used to assess the pulse in the foot

5. _____ Artery that extends from the common iliac artery toward the thigh

6. _____ Branch of the abdominal aorta that supplies the kidneys

7. _____ Artery that extends from the subclavian artery toward the arm

8. _____ Artery that arises directly from the arch of the aorta, travels up the anterolateral part of the neck, and supplies the brain

9. _____ Artery used to assess blood pressure in the arm

10. _____ Largest artery in the body; all other arteries emerge from it directly or indirectly.

11. _____ Artery in the forearm and located on the "thumb side"; used to assess the pulse at the wrist

12. _____ Popliteal artery

13. _____ This "trunk" emerges from the abdominal aorta; it branches into the hepatic, gastric, and splenic arteries.

14. _____ Superior mesenteric artery

15. _____ An extension of the external iliac artery; it runs through the thigh and extends into the leg.

16. _____ Inferior mesenteric artery

17. _____ An extension of the common carotid artery that supplies the brain

18. _____ Artery located in the forearm on the "little finger" side

19. _____ Left common carotid artery

20. _____ Posterior tibial artery

21. _____ Arises from the aortic arch and extends to the shoulder as the left axillary artery; gives rise to the left vertebral artery

22. _____ Supplies the anterior leg

23. _____ Arises from the aortic arch and bifurcates into the right common carotid artery and the right subclavian artery

24. _____ A branch of the common iliac artery that supplies the anterior thigh region and the organs within the pelvis

Veins

Directions: Referring to the figure, write the correct number in the space provided. Some numbers may be used more than once. See text, pp. 349-353.

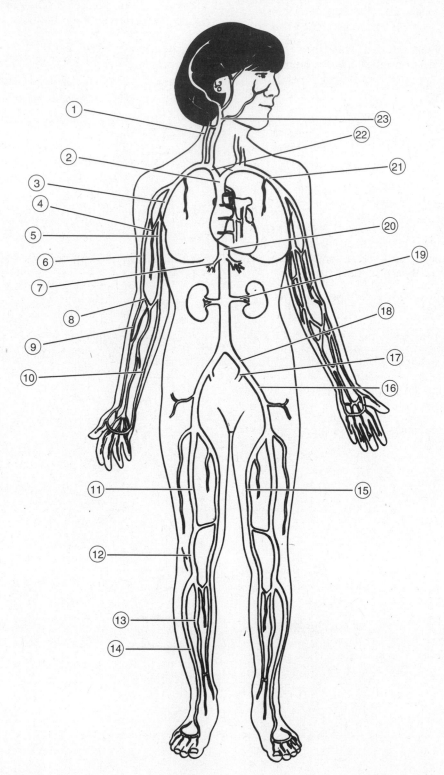

1. _____, _____ Two veins that form the venae cavae

2. _____ "Main vein that drains the brain"

3. _____ The vein that empties directly into the vena cava from the thigh

4. _____, _____ The veins in the thigh that drain directly into the common iliac vein

5. _____ Longest vein in the body; it begins in the foot, ascends medially, and merges with the femoral vein to become the external iliac vein

6. _____ Deep vein that runs through the thigh; it enters the pelvis as the external iliac vein

7. _____ Vein that receives blood from the kidneys and drains into the inferior vena cava

8. _____, _____ Deep veins of the leg that drain the foot and leg; they merge to form the popliteal vein

9. _____ Popliteal vein

10. _____ Median cubital vein

11. _____ Located in the armpit region; receives unoxygenated blood from the brachial, basilic, and cephalic veins

12. _____ Drains the liver, emptying into the inferior vena cava

13. _____ Vein into which the jugular vein empties

14. _____ The subclavian vein continues toward the heart as this vein.

15. _____ The external jugular vein

MATCHING

Arteries

Directions: Match the following terms to the most appropriate definition by writing the correct letter in the space provided. Some terms may be used more than once. See text, pp. 346-349.

A. celiac trunk
B. circle of Willis
C. coronary arteries
D. renal artery
H. dorsalis pedis artery
I. aorta
J. basilar artery
K. hepatic artery

E. internal carotid arteries
F. common iliac arteries
G. mesenteric arteries
L. left subclavian artery

1. _____ The vertebral arteries pass upward from the subclavian arteries toward the back of the neck; they extend into the cranium and join to form this artery.

2. _____ A branch of the celiac trunk that supplies the liver

3. _____ Largest artery in the body; arises from the left ventricle of the heart

4. _____ Artery that supplies the kidney

5. _____ These arteries are branches of the abdominal aorta; they supply blood to most of the small intestine and part of the large intestine.

6. _____ An arrangement of arterial blood vessels found at the base of the brain

7. _____ An extension of the anterior tibial artery that supplies the foot

8. _____ The distal end of the abdominal aorta splits, or bifurcates, into these.

9. _____ Short artery that divides into the gastric artery, splenic artery, and hepatic artery

10. _____ Large artery that is classified as ascending, arch, and descending

11. _____ Large artery that is classified as thoracic and abdominal

12. _____ Branches of the ascending aorta that supply the myocardium of the heart

13. _____ Arteries that ascend on the anterolateral aspect of the neck; supply the circle of Willis

14. _____ Branch of the aortic arch that supplies the left shoulder and upper arm

15. _____ Name that means *heavy sleep* or *stupor*

MATCHING

Veins

Directions: Match the following terms to the most appropriate definition by writing the correct letter in the space provided. Some terms may be used more than once. See text, pp. 349-351.

A. superior vena cava
B. inferior vena cava
C. jugular veins
D. hepatic veins
E. renal vein
F. subclavian vein
G. portal vein
H. great saphenous vein
I. median cubital vein
J. femoral vein

1. _____ Located in the lower extremity; the longest vein in the body

2. _____ Large deep vein in the thigh that enters the pelvis as the external iliac vein

3. _____ The common iliac vein continues as this vein.

4. _____ This large vein drains the head, shoulders, and upper extremities; it empties into the right atrium.

5. _____ Formed from the union of the superior mesenteric vein and the splenic vein

6. _____ Drain the head

7. _____ Large vein that returns blood to the right atrium from all the regions below the diaphragm

8. _____ Drains blood from the kidney and empties it into the inferior vena cava

9. _____ Large vein that carries blood from the digestive organs to the liver

10. _____ Drains the liver and empties blood into the inferior vena cava

11. _____ Receives blood from the axillary vein and the external jugular vein

12. _____ Long superficial vein often "borrowed" for cardiac bypass surgery

13. _____ Vein that joins the cephalic and basilic veins

14. _____ Arm vein that is commonly used to withdraw a sample of blood

Hepatic Portal System

Directions: Referring to the figure, write the correct number in the spaces on the next page. Some numbers may be used more than once. See text, pp. 353, 354.

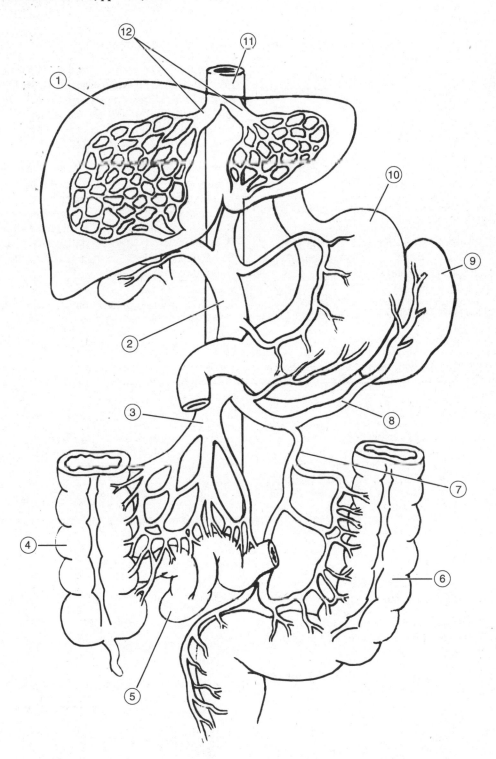

1. _____ Large vein that is formed by the merger of the superior mesenteric and splenic veins

2. _____ Vein that drains most of the small intestine and part of the large intestine (ascending colon)

3. _____ Descending colon

4. _____ Veins that bring blood from the liver to the inferior vena cava

5. _____ Stomach

6. _____ Spleen

7. _____ Liver

8. _____ Vein that receives blood from the inferior mesenteric vein

9. _____ Superior mesenteric vein

10. _____ Small intestine

11. _____ Part of large intestine; ascending colon

12. _____ Inferior vena cava

MATCHING

Fetal Circulation

Directions: Match the following terms to the most appropriate definition by writing the correct letter in the space provided. Some terms may be used more than once. See text, pp. 353-356.

A. foramen ovale   D. umbilical cord
B. ductus venosus   E. placenta
C. ductus arteriosus

1. _____ Structure that carries three blood vessels (two arteries and one vein); connects the mother with the fetus

2. _____ Hole in the interatrial septum that allows fetal blood to shunt from the right atrium to the left atrium

3. _____ Place for maternal-fetal exchange of nutrients, gases, and waste

4. _____ Structure that connects the umbilical vein with the fetal inferior vena cava; partly shunts blood past the fetal liver

5. _____ Short tube that connects the fetal pulmonary artery with the thoracic aorta

6. _____ Structure that serves as lungs for the fetus

7. _____ Failure of this opening to close after birth creates a left-to-right shunt between the aorta and pulmonary artery.

8. _____ Failure of this opening to close after birth creates a left-to-right shunt between the atria.

TRACE A DROP OF BLOOD

Directions: Refer to Figs. 18.6 and 18.7 in the textbook (pp. 350 and 352), and using the words below, fill in the blanks. The words can be used more than once (L is left, R is right). You will also need the information about blood flow through the heart in Chapter 16 (Figs. 16.4, 16.6, and 16.7 on pp. 312, 316, and 317).

circumflex artery   inferior vena cava
L atrium   R common iliac vein
R axillary vein   superior vena cava
abdominal aorta   L femoral artery
R subclavian artery   R brachial artery
R atrium   pulmonary artery
R popliteal vein   R brachiocephalic vein
L femoral vein

1. Pulmonary artery → pulmonary capillaries → pulmonary veins → _____ → left ventricle → ascending aorta → aortic arch → right brachiocephalic artery → _____ → right axillary artery → _____ → right radial artery

2. Thoracic aorta → _____ → left common iliac artery → left external iliac artery → _____ → left popliteal artery → left anterior tibial artery → left dorsalis pedis artery

3. Right internal jugular vein → right subclavian vein → _____ → superior vena cava → right atrium → right ventricle → _____ → pulmonary capillaries → left atrium

4. Right radial vein → right brachial vein → _____ → right subclavian vein → right brachiocephalic vein → _____ → right atrium

5. Right anterior and posterior tibial veins → _____ → right femoral vein → right external iliac → _____ → inferior vena cava → right atrium

6. Left saphenous vein → _____ → left external iliac vein → left common iliac vein → _____ → right atrium

7. Ascending aorta → left coronary artery → _____ → walls of the left side of the heart → cardiac veins → cardiac sinus → _____

TRACE A DROP OF BLOOD, AGAIN

Directions: Refer to Figs. 18.8, 18.9, and 18.10 in the textbook (pp. 354, 355, and 356), and using the words below, fill in the blanks. The words can be used more than once. Also see Figs. 16.4 and 16.6 (pp. 312 and 316).

umbilical vein	pulmonary veins
inferior vena cava	portal vein
foramen ovale	umbilical arteries
circle of Willis	hepatic vein
L common carotid artery	R atrium
R ventricle	pulmonary artery
R common carotid artery	superior mesenteric vein
ductus arteriosus	

1. _____ → ductus venosus → inferior vena cava → right atrium → _____ → left atrium → left ventricle → ascending aorta → descending aorta → _____ → placenta

2. _____ → ductus venosus → _____ → right atrium → right ventricle → pulmonary artery → _____ → thoracic aorta → abdominal aorta → internal iliac arteries → _____ → placenta

3. Pulmonary capillaries → _____ → left atrium → left ventricle → ascending aorta → aortic arch → _____ → left internal carotid artery → _____

4. Ascending aorta → aortic arch → right brachiocephalic artery → _____ → right internal carotid artery → _____

5. Splenic vein → _____ → hepatic sinusoids → hepatic veins → _____ → right atrium → _____ → pulmonary artery

6. _____ → portal vein → hepatic sinusoids → _____ → inferior vena cava → _____ → right ventricle → _____ → pulmonary capillaries → pulmonary veins → left atrium → left ventricle → aorta

GETTING FROM POINT A TO POINT B

Directions: For each of the following, indicate the blood vessel or structure NOT used as blood flows from Point A to Point B.

1. Point A: aortic arch → Point B: dorsalis pedis artery
   a. saphenous
   b. common iliac artery
   c. descending aorta
   d. femoral artery

2. Point A: left atrium → Point B: circle of Willis
   a. internal carotid artery
   b. aortic valve
   c. ascending aorta
   d. portal vein

3. Point A: heart → Point B: foot
   a. common carotid artery
   b. thoracic aorta
   c. internal iliac artery
   d. descending aorta

4. Point A: foot → Point B: heart
   a. saphenous
   b. inferior vena cava
   c. jugular
   d. popliteal vein

5. Point A: pulmonary veins → Point B: kidney
   a. pulmonic valve
   b. left atrium
   c. thoracic aorta
   d. renal artery

6. Point A: femoral vein → Point B: pulmonary artery
   a. subclavian vein
   b. inferior vena cava
   c. tricuspid valve
   d. external iliac vein

7. Point A: median cubital vein → Point B: superior vena cava
   a. axillary vein
   b. brachiocephalic vein
   c. subclavian vein
   d. circle of Willis

8. Point A: splenic vein → Point B: inferior vena cava
   a. hepatic vein
   b. hepatic sinusoids
   c. celiac trunk
   d. portal vein

9. Point A: subclavian vein → Point B: descending aorta
   a. aortic valve
   b. inferior vena cava
   c. ascending aorta
   d. pulmonary veins

10. Point A: pulmonary → Point B: circle
    artery   of Willis
    a. pulmonary veins   c. jugular
    b. internal carotid artery   d. aortic arch

11. Point A: brachiocephalic → Point B: pulmonic
    vein   valve
    a. superior vena cava   c. right atrium
    b. jugular   d. tricuspid valve

12. Point A: umbilical vein → Point B: left atrium
    a. ductus venosus   c. inferior vena cava
    b. ductus arteriosus   d. foramen ovale

13. Point A: jugular → Point B: pulmonary capillaries
    a. ascending aorta   c. pulmonary artery
    b. superior vena cava   d. tricuspid valve

14. Point A: →   Point B: left anterior
    pulmonary capillary   descending
    coronary artery
    a. aortic valve   c. arch of the aorta
    b. ascending aorta   d. left atrium

15. Point A: right atrium → Point B: ductus arteriosus
    a. pulmonary artery   c. pulmonary veins
    b. pulmonic valve   d. right ventricle

16. Point A: ductus → Point B: umbilical
    arteriosus   vein
    a. descending aorta   c. circle of Willis
    b. umbilical arteries   d. placenta

17. Point A: right ventricle → Point B: renal artery
    a. jugular   c. thoracic aorta
    b. arch of the aorta   d. pulmonary capillaries

18. Point A: superior → Point B: pulmonic
    mesenteric vein   valve
    a. right atrium   c. jugular
    b. portal   d. hepatic vein

19. Point A: ascending → Point B: circle of
    aorta   Willis
    a. brachiocephalic artery   c. jugular
    b. basilar artery   d. vertebral artery

20. Point A: posterior → Point B: pulmonary
    tibial vein   artery
    a. pulmonary veins   c. inferior vena cava
    b. tricuspid valve   d. saphenous

COLORING: RED AND BLUE

*Directions: The following blood vessels carry oxygen-
ated blood or unoxygenated blood. Color the box blue if
the blood vessels carry unoxygenated blood. Color the
box red if the blood vessels carry oxygenated blood.*

1. ❏ pulmonary veins

2. ❏ aorta

3. ❏ portal

4. ❏ umbilical vein

5. ❏ venae cavae

6. ❏ jugular

7. ❏ pulmonary artery

8. ❏ celiac trunk

9. ❏ great saphenous

10. ❏ internal carotid artery

11. ❏ dorsalis pedis artery

12. ❏ circle of Willis

13. ❏ ductus venosus

14. ❏ umbilical arteries

15. ❏ hepatic veins

16. ❏ coronary artery

17. ❏ subclavian artery

18. ❏ common iliac artery

19. ❏ superior mesenteric vein

20. ❏ brachial artery

Chapter **18 Anatomy of the Blood Vessels**

SIMILARS AND DISSIMILARS

Directions: Circle the word in each group that is least similar to the others. Indicate the similarity of the three words on the line below each question.

1. intima   portal   adventitia   media

   _____

2. arch   descending   common carotid   ascending

   _____

3. conductance   arteriole   resistance   exchange

   _____

4. brachial   radial artery   dorsalis   axillary artery
   artery   pedis

   _____

5. internal carotid   popliteal artery

   femoral artery   anterior tibial artery

   _____

6. common   vertebral   portal   circle of Willis
   carotid

   _____

7. portal   superior   splenic vein   common
             mesenteric vein   carotid

   _____

8. great   jugular   femoral   posterior
   saphenous   vein   tibial vein

   _____

9. pulmonary   aorta   superior   inferior
   artery   vena cava   vena cava

   _____

10. ductus   foramen   circle of   ductus
    arteriosus   ovale   Willis   venosus

    _____

11. umbilical   gastric   splenic   hepatic
    arteries   artery   artery   artery

    _____

12. foramen   right atrium   left ventricle   left
    ovale   atrium

    _____

13. venae cavae   aorta   dorsalis pedis   pulmonary
                                              artery

    _____

14. ductus   foramen ovale   circle of   umbilical
    venosus   Willis   vein

    _____

15. basilic   cephalic   median cubital   aorta

    _____

16. varicosity   hemorrhoids

    anastomoses   esophageal varices

    _____

17. jugular   basilar   internal carotid   circle of
                                              Willis

    _____

Part II: Putting It All Together

MULTIPLE CHOICE

Directions: Choose the correct answer.

1. Blood that is a bright red color (as opposed to a bluish-red color)
   a. is always found in arteries.
   b. is never found in veins.
   c. is oxygenated.
   d. has a high concentration of carbaminohemoglobin.

2. About 70% of the blood volume is located in which structures?
   a. veins
   b. capillaries
   c. resistance vessels
   d. arteries

3. Arterioles have a lot of smooth muscle that allows them to
   a. act as exchange vessels.
   b. store large amounts of blood.
   c. contract and relax, thereby affecting blood vessel diameter.
   d. prevent a backward flow of blood.

153

4. The internal carotid and vertebral arteries
   a. are called the *circle of Willis.*
   b. empty blood into the subclavian arteries.
   c. supply oxygenated blood to the arteries of the brain.
   d. drain cerebrospinal fluid from the brain.

5. The purpose of the ductus arteriosus is to
   a. oxygenate blood.
   b. bypass the fetal lungs.
   c. bypass the fetal liver.
   d. conduct blood to the fetal pulmonary capillaries.

6. The foramen ovale
   a. is found in the interventricular septum.
   b. bypasses the fetal liver.
   c. shunts blood from the pulmonary artery to the aorta.
   d. shunts blood from the right side of the heart to the left side.

7. Failure of the ductus arteriosus to close after birth
   a. causes immediate death.
   b. decreases blood flow to the lungs.
   c. causes blood to shunt from the pulmonary artery to the aorta.
   d. causes a left-to-right shunt.

8. Which of the following is a correct statement about the fetal circulation?
   a. There is one umbilical artery.
   b. There are two umbilical veins.
   c. The umbilical arteries carry oxygenated blood.
   d. The umbilical vein carries oxygenated blood.

9. The jugulars
   a. carry unoxygenated blood from the brain.
   b. are part of the hepatic portal circulation.
   c. are exchange blood vessels.
   d. are resistance vessels.

10. Most often, atherosclerosis of a carotid artery causes
    a. facial flushing.
    b. cyanosis and breathlessness.
    c. cognitive impairment.
    d. cold, pale extremities.

11. Right-sided heart failure is most likely to cause
    a. pulmonary edema.
    b. cyanosis and orthopnea.
    c. jugular vein distention.
    d. cough, air hunger, and restlessness.

12. The portal vein
    a. supplies most of the oxygenated blood to the liver.
    b. carries blood from the digestive organs directly to the inferior vena cava.
    c. carries blood that is rich in digestive end products to the liver.
    d. receives blood from the hepatic veins.

13. The superior mesenteric and splenic veins
    a. receive unoxygenated blood from the portal vein.
    b. bypass the liver and empty blood into the inferior vena cava.
    c. merge to form the portal vein.
    d. deliver blood to the stomach and intestines to aid in digestion.

14. To what blood vessel do the follow words refer: ascending, arch, descending, thoracic, and abdominal?
    a. venae cavae
    b. aorta
    c. circle of Willis
    d. hepatic-portal circulation

15. The subclavian veins
    a. are found only in the fetal circulation.
    b. are part of the hepatic portal circulation.
    c. are part of the circle of Willis.
    d. empty blood into the brachiocephalic veins.

16. What is true of both the anterior tibial artery and the dorsalis pedis artery?
    a. Both run parallel to the great saphenous vein.
    b. Both are part of the hepatic portal circulation.
    c. Both are found in the lower extremities.
    d. Both contain valves.

17. The basilar artery
    a. delivers blood to the circle of Willis.
    b. is part of the celiac trunk.
    c. supplies oxygenated blood to the liver.
    d. is a branch of the aortic arch that becomes the vertebral artery as it ascends the posterior neck region.

18. Which of the following is correct?
    a. abdominal aorta→thoracic aorta→coronary arteries
    b. splenic vein→hepatic veins→portal vein
    c. descending aorta→common iliac arteries→great saphenous vein→dorsalis pedis artery
    d. subclavian vein→brachiocephalic vein→superior vena cava→right atrium

19. The coronary arteries
    a. arise from the pulmonary artery.
    b. arise from the ascending aorta.
    c. deliver blood to the pulmonary capillaries.
    d. deliver blood to the circle of Willis.

20. A drop of blood flows from the superior mesenteric vein to the right atrium. Through which blood vessel must it flow?
    a. great saphenous vein
    b. dorsalis pedis artery
    c. portal vein
    d. jugular vein

21. A drop of blood flows from the ascending aorta to the basilar artery. Through which blood vessel must it flow?
    a. portal vein
    b. renal artery
    c. vertebral artery
    d. external carotid artery

22. Which of the following best describes an artery? A blood vessel that
    a. always carries oxygenated blood.
    b. is always color-coded red.
    c. contains valves ensuring the flow of blood from the arterial to the venous side of the circulation.
    d. carries blood away from the heart.

23. What type of blood vessel is described as follows: carries blood toward the heart, called capacitance vessels, are valve-containing?
    a. arteriole
    b. capillary
    c. vein
    d. sinusoid

24. Which of the following is true of both the aortic arch and the right brachiocephalic artery? Both
    a. arise from the coronary arteries.
    b. deliver blood to the jugulars.
    c. receive blood from the circle of Willis.
    d. give rise to the common carotid arteries.

25. Which of the following is true of the renal arteries?
    a. They drain urine from the kidneys.
    b. They emerge from the abdominal aorta and supply the kidneys with oxygenated blood.
    c. They form the anterior blood vessels that form the circle of Willis.
    d. They are direct branches of the common iliac arteries.

CASE STUDY

Polly Phagia is a type 2 diabetic with neuropathy and atherosclerosis of the blood vessels of the lower extremities. She is admitted to your unit because of an area of her great toe that appears blackened.

1. What is the concern (if any) of the blackened toe?
    a. There is no concern; we can assume that she stubbed her toe.
    b. The poor circulation starves the tissue of oxygen and nutrients, causing gangrene.
    c. She must be in excruciating pain.
    d. She has lost her sense of balance.

2. What is the effect of poor arterial circulation to the feet (as is true with many diabetics)?
    a. The feet feel warm.
    b. The dorsalis pedis pulse may be undetectable.
    c. The feet sweat excessively.
    d. The feet appear red.

PUZZLE

Hint: Vain Jane Complains of Vein Pain

Directions: Perform the following functions on the Sequence of Words that follows. When all the functions have been performed, you are left with a word or words related to the hint. Record your answer in the space provided.

Functions: Remove the following:

1. Main vein that drains the brain

2. Main vein named superior and inferior

3. Longest vein in the leg

4. Large deep vein in the thigh

5. Large vein that delivers blood to the liver

6. Vein that merges with the superior mesenteric vein to form the portal vein

7. Vein that drains the jugular vein

8. Chamber of the heart that receives blood from the venae cavae

9. Veins that carry oxygenated blood from the lungs to the left side of the heart

10. A vein in the arm

11. Vein that carries oxygenated blood from the placenta to the fetus

12. Veins that carry blood from the liver toward the inferior vena cava

13. Vein that drains blood from the kidney and delivers it to the inferior vena cava

14. Veins that merge to form the distal inferior vena cava

Sequence of Words

V E N A E C A V A E U M B I L I C A L J U G U L A R
S P L E N I C P O R T A L S U B C L A V I A N T H R O
M B O P H L E B I T I S F E M O R A L H E P A T I C S
A P H E N O U S P U L M O N A R Y R E N A L B A S
I L I C R I G H T A T R I U M C O M M O N I L I A C S

Answer: _____

19 Functions of the Blood Vessels

Answer Key: Textbook page references are provided as a guide for answering these questions. A complete answer key was provided for your instructor.

OBJECTIVES

1. List the five functions of the blood vessels.
2. Discuss blood pressure, including:
   - Describe the measurement of blood pressure.
   - Explain the variance of blood pressure in different blood vessels.
   - Describe the factors that determine blood pressure.
   - Explain the mechanisms involved in regulation of blood pressure, including the baroreceptor reflex.
3. Explain how blood vessels act as exchange vessels, including:
   - Describe the factors that determine capillary exchange.
   - Describe the mechanisms of edema formation.
4. Explain how the blood vessels respond to changing body needs.
5. Describe the role of the blood vessels in the regulation of body temperature.

Part I: Mastering the Basics

MATCHING

Directions: Use the following words to fill in the blanks. Some words are used more than once. See text, pp. 361-370.

A. capillaries	E. capillary filtration pressure
B. blood pressure	F. arterioles
C. albumin	G. ischemia
D. edema	

1. _____ Resistance vessels

2. _____ Exchange vessels

3. _____ Determines systemic vascular resistance (SVR)

4. _____ Pulselessness, pallor, pain, cool to touch, and gangrene

5. _____ Determined by cardiac output × SVR

6. _____ Determines plasma oncotic pressure

7. _____ Impaired blood flow to a tissue or organ

8. _____ A deficiency of this plasma protein causes edema

9. _____ Determined by stroke volume × heart rate × SVR

10. _____ Abnormal collection of fluid

11. _____ The force exerted by the blood against the walls of the blood vessels

12. _____ The outward pushing pressure within the exchange vessels

MATCHING

Blood Pressure

Directions: Match the following terms to the most appropriate definition by writing the correct letter in the space provided. Some terms may be used more than once. See text, pp. 361-365.

A. hypotension	H. pulse
B. normal blood pressure	I. systolic pressure
C. sphygmomanometer	J. vasopressors
D. pulse pressure	K. postural hypotension
E. hypertension	L. diastolic pressure
F. systemic vascular resistance (SVR)	M. Korotkoff sounds
G. brachial	N. mean arterial blood pressure (MABP)

1. _____ A sudden drop in blood pressure when the person moves from a lying to a standing position; often causes dizziness and fainting

2. _____ Vibrations of the blood vessel walls that reflect heart rate

3. _____ The noise heard through the stethoscope when blood pressure is measured

4. _____ In addition to cardiac output, this determines blood pressure.

5. _____ The highest pressure recorded during systole

6. _____ The artery most commonly used to measure blood pressure

7. _____ The difference between the systolic reading and the diastolic reading

8. _____ This is most likely to be caused by a massive peripheral vasodilation.

9. _____ Diastolic blood pressure + one-third pulse pressure

10. _____ Blood pressure of 116/72 mm Hg

11. _____ Hormones or drugs that increase blood pressure

12. _____ This is most likely to be caused by intense peripheral vasoconstriction.

13. _____ A device used to measure blood pressure

14. _____ Blood pressure reading of 160/95 mm Hg

15. _____ Blood pressure reading of 75/40 mm Hg—a "shocky" blood pressure

16. _____ The pressure in the arteries during ventricular relaxation

17. _____ Silence . . . tap . . . tap . . . tap . . . muffle

18. _____ The top number of a blood pressure reading

19. _____ The bottom number of a blood pressure reading

READ THE DIAGRAM

Directions: Referring to the illustration, fill in the spaces on the next page with the correct numbers.

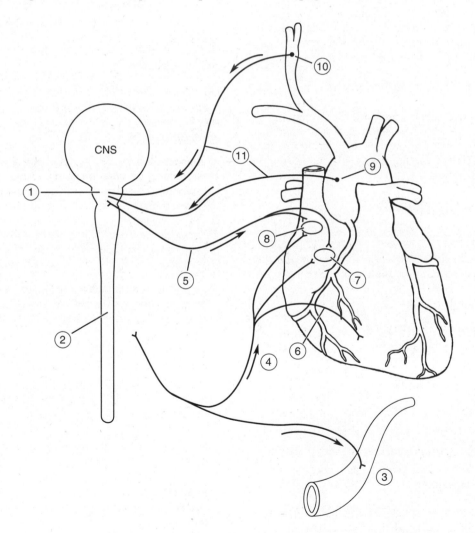

1. _____ Sympathetic nerve stimulation causes vasoconstriction.

2. _____ Sympathetic nerve stimulation increases heart rate.

3. _____ Autonomic stimulation is interpreted in the medulla oblongata.

4. _____ Carotid baroreceptors

5. _____ Sympathetic stimulation causes a (+) chronotropic and (+) inotropic effect.

6. _____ CNs IX and X carry baroreceptor information to the medulla oblongata.

7. _____ Determines SVR

8. _____ Vagal stimulation decreases heart rate.

9. _____ Sympathetic stimulation increases force of myocardial contraction.

10. _____ Aortic baroreceptors

11. _____ Vagal discharge affects the sinoatrial (SA) and atrioventricular (AV) nodal activity.

12. _____ These motor fibers are part of the craniosacral outflow.

13. _____ The fibers are part of the thoracolumbar outflow.

14. _____ SA node; pacemaker

15. _____ AV node

16. _____ Origin of the sympathetic nerves that transmit response to the heart and arterioles

MATCHING

Blood Pressure Readings

Directions: Refer to the following blood pressure readings and fill in the blanks below. See text, pp. 358-359.

A. 115/75 mm Hg   C. 72/48 mm Hg
B. 210/105 mm Hg

1. _____ Blood pressure that causes a reflex tachycardia

2. _____ Blood pressure that may require a vasopressor drug

3. _____ Normotension

4. _____ Blood pressure that may require an antihypertensive drug such as a calcium channel blocker

5. _____ Blood pressure that may require the use of an alpha$_1$-adrenergic antagonist (blocker)

6. _____ Blood pressure caused by hemorrhage

7. _____ Blood pressure that may result from excess catecholamines such as epinephrine

8. _____ Blood pressure that may be caused by intense sympathetic discharge

9. _____ This person has a pulse pressure of 105 mm Hg.

10. _____ This person has a mean arterial blood pressure (MABP) of 56 mm Hg.

11. _____ This person has a pulse pressure of 40.

12. _____ This person has an MABP of 89 mm Hg.

13. _____ Blood pressure that is most likely to cause a stroke or brain attack

14. _____ Blood pressure that is most likely to underperfuse the heart muscle, causing chest pain and possibly a myocardial infarction (MI)

15. _____ This blood pressure is most likely to be treated with a drug that causes peripheral vasodilation.

ORDERING

Directions: Referring to the two case studies, order the events that occur when a sudden change in blood pressure activates the baroreceptor response. Place the events listed below in the spaces provided.

Case 1. Mr. C has a 10-year history of hypertension, for which he was being treated with an antihypertensive drug. His physician recently prescribed an additional antihypertensive drug. On the first day that he took both drugs, Mr. C experienced the following. On arising from a sitting position, he became extremely dizzy and indicated that his heart was racing.

1. The medulla oblongata interprets the sensory information and fires the sympathetic nervous system.

2. Information is carried from the baroreceptors by the sensory nerves to the medulla oblongata.

3. Mr. C experienced the reflex tachycardia as a "racing" heart.

4. Sympathetic nerve stimulation increased heart rate, stroke volume, and SVR, thereby increasing blood pressure.

5. The blood pressure suddenly declined in response to the combined effects of the drugs.

Chapter **19** Functions of the Blood Vessels

6. The sudden decline in blood pressure activated the carotid and aortic baroreceptors.

1. _____

2. _____

3. _____

4. _____

5. _____

6. _____

Case 2. Ms. G was showering using a new pulsatile shower head. With no warning, she experienced an episode of syncope while showering. She was subsequently diagnosed with a carotid sinus syncope.

1. The medulla oblongata interprets the sensory information falsely as an increased blood pressure and fires the parasympathetic nervous system.

2. Information is carried from the baroreceptors by the sensory nerves to the medulla oblongata.

3. The low blood pressure decreased blood flow to the brain, and Ms. G fainted.

4. Parasympathetic nerve stimulation (and diminished sympathetic activity) decreased heart rate and SVR, thereby decreasing blood pressure.

5. The sudden "increase" in blood pressure activated the carotid and aortic baroreceptors.

1. _____

2. _____

3. _____

4. _____

5. _____

MATCHING

Nervous Stimulation of the Heart and Blood Vessels

Directions: Indicate whether the following are caused by the firing of sympathetic nerves (S) or parasympathetic nerves (P) by writing the correct letter in the space provided. See text, pp. 365-367.

1. _____ Decrease in heart rate

2. _____ Same as a vagal discharge

3. _____ Positive (+) inotropic effect

4. _____ Increased blood pressure

5. _____ Increased force of myocardial contraction

6. _____ Increased cardiac output

7. _____ Increased SVR

8. _____ (+) Chronotropic effect

9. _____ Negative (–) dromotropic effect

10. _____ Most likely to cause heart block

11. _____ Most likely to cause tachycardia

12. _____ Most likely to cause bradycardia

13. _____ Mimics the effects of the catecholamines

SIMILARS AND DISSIMILARS

Directions: Circle the word in each group that is least similar to the others. Indicate the similarity of the three words on the line below each question.

1. pain   pallor   oncotic pressure   pulselessness

_____

2. baroreceptor   edema   carotid sinus   aortic arch

_____

3. albumin   colloidal oncotic   Korotkoff   edema
                pressure   sounds

_____

4. blood pressure   edema   cardiac output   SVR

_____

5. ischemia   tissue   gangrene   oncotic pressure
    hypoxia

_____

6. 120/80 mm Hg   <u>systolic</u>   pulse   blood
                   diastolic   pressure

_____

MULTIPLE CHOICE

Directions: Choose the correct answer.

1. Reflex tachycardia is most likely to develop in response to
   a. pedal edema.
   b. blunting of the baroreceptor reflex by a $beta_1$-adrenergic blocker.
   c. sudden decline in blood pressure.
   d. expanded blood volume.

2. A person with poor skin turgor is most likely to have this condition.
   a. edema
   b. jaundice
   c. cyanosis
   d. dehydration

3. Which condition is most likely to be caused by an increase in SVR?
   a. increased heart rate
   b. increased blood pressure
   c. hypotension
   d. anemia

4. A vasopressor substance
   a. thins the blood.
   b. decreases heart rate.
   c. decreases cardiac output.
   d. increases blood pressure.

5. What is the stimulus for the baroreceptor reflex?
   a. change in blood pH
   b. heart rate
   c. oxygen saturation of blood in the aortic arch and carotid sinus
   d. stretch of receptors located in several of the large arteries

6. What is the effect of stimulation of the sympathetic nerve?
   a. peripheral vasodilation
   b. decrease in cardiac output
   c. decrease in heart rate
   d. increase in blood pressure

7. Which of the following is regulated by the renin-angiotensin-aldosterone system?
   a. blood pressure
   b. plasma protein synthesis
   c. plasma levels of calcium
   d. blood glucose level

8. During strenuous exercise, the percentage of blood flow to which structure greatly increases?
   a. kidney
   b. digestive tract
   c. skeletal muscles
   d. reproductive system

9. Which condition occurs if the amount of water filtered out of the capillary exceeds the amount of water reabsorbed into the capillary?
   a. dehydration
   b. poor skin turgor
   c. edema
   d. inflammation

10. Which of the following contributes to an increased venous return of blood during exercise?
    a. increased capillary filtration pressure
    b. increased oncotic pressure
    c. effect of venoconstriction and the skeletal muscle pump
    d. activation of erythropoiesis by the bone marrow

11. Which measurement do you need to calculate pulse pressure?
    a. hematocrit, 45%
    b. heart rate, 72 beats/min
    c. pulse deficit, 5 beats
    d. blood pressure, 120/80 mm Hg

12. Which measurement do you need to calculate mean arterial pressure?
    a. pulse pressure, 40 mm Hg
    b. heart rate, 72 beats/min
    c. pulse deficit, 5 beats
    d. blood pressure, 120/80 mm Hg

13. Which of the following is least true of vasoconstriction?
    a. caused by sympathetic discharge
    b. decreases cardiac output and blood pressure
    c. increases afterload
    d. moves blood from the venous circulation to the arterial circulation

161

14. A person has a history of essential hypertension; he has been taking flu medication for the past 3 weeks. The flu medicine contains phenylephrine, an alpha$_1$-adrenergic agonist. Which adverse effect might the person experience?
    a. stuffy nose from congested nasal mucous membrane
    b. further elevation in blood pressure
    c. drowsiness and fatigue
    d. constipation and bloating

15. Which of the following cannot be determined from a blood pressure reading of 130/72 mm Hg?
    a. systolic reading
    b. diastolic reading
    c. pulse
    d. pulse pressure

16. Which of the following makes the capillaries ideal exchange vessels?
    a. large area for diffusion
    b. slow velocity of blood flow
    c. thin capillary membrane
    d. all of the above

17. A collar that exerts pressure over the baroreceptors may
    a. cause hypotension and fainting.
    b. cause a hypertensive crisis and stroke.
    c. cause cerebral ischemia and loss of consciousness.
    d. inhibit venous drainage, causing jugular vein distention.

18. Plasma oncotic pressure
    a. pulls water into the capillary from the interstitium.
    b. pushes water out of the capillary into the interstitium.
    c. pushes water from the interstitium into the cells.
    d. pumps K^+ into the cells.

19. What is the effect of an age-related stiffening of the blood vessel walls?
    a. anemia
    b. increased blood pressure
    c. increased cardiac output
    d. decreased diastolic pressure

20. Why does blood flow from the arterial side of the circulation to the venous side of the circulation?
    a. Capillary pressure is higher than either arterial or venous pressure.
    b. The oncotic pressure in the veins is greater than in the arteries.
    c. Pressure on the arterial side of the circulation is greater than on the venous side of the circulation.
    d. Blood volume is greater in the arterial circulation than in the venous circulation.

21. When the capillary filtration (hydrostatic) pressure exceeds plasma oncotic pressure water moves from the
    a. interstitium into the capillaries.
    b. capillaries into the interstitium.
    c. lymphatic vessels into the interstitium.
    d. interstitium into the lymphatic vessels.

22. Postural hypotension is most apt to cause
    a. white coat hypertension.
    b. syncope (fainting).
    c. widened pulse pressure.
    d. bradycardia.

23. Which of the following is true?
    a. CO = ejection fraction × SV
    b. BP = CO × cardiac reserve
    c. BP = HR × SV × SVR
    d. BP = SV × preload

CASE STUDY

C.L. went to the physician for his annual physical. Although he had no symptoms, his blood pressure was elevated at 160/95 mm Hg. He was advised to lose weight, begin an exercise program, and take medication for his blood pressure.

1. What is the goal of therapy?
    a. decrease the systolic reading to 90 mm Hg
    b. decrease the pulse pressure to 10
    c. decrease both the systolic and diastolic pressures
    d. achieve a heart rate of 60 beats/min

2. Which type of drug acts as an antihypertensive agent?
    a. one that stimulates the sympathetic nerves
    b. one that increases peripheral resistance
    c. one that causes a vasopressor effect
    d. one that causes vasodilation, thereby decreasing SVR

Part III: Challenge Yourself!

PUZZLE

Hint: Why Is Al Bumin a Swell Fellow?

Directions: Perform the following functions on the Sequence of Words that follows. When all the functions have been performed, you are left with a word or words related to the hint. Record your answer in the space provided.

Functions: Remove the following:

1. Consequence of diminished blood flow to a tissue

2. Top and bottom blood pressure numbers

3. Name of the resistance vessels and exchange vessels

4. Difference between the systolic and diastolic blood pressures

5. Three determinants of blood pressure

6. Effects of straining at stools

7. Location (two) of baroreceptors

8. Sounds heard while assessing blood pressure: tap tap tap . . . .

9. Higher than normal blood pressure; lower than normal blood pressure

10. Responsible for capillary oncotic pressure

11. Two changes in blood vessel diameter that affect SVR

Sequence of Words

CAPILLARIESHYPOTENSIONCARO
TIDSINUSVASOCONSTRICTIONON
COTIC(SWELLING)PRESSURESYST
OLICPRESSUREKOROTKOFFSOUN
DSDIASTOLICPRESSUREHEARTRA
TEPLASMAPROTEINSVALSALVAM
ANEUVERARTERIOLESSTROKEVOL
UMEVASODILATIONAORTICARCH
HYPERTENSIONPULSEPRESSUREI
SCHEMIASYSTEMICVASCULARRE
SISTANCE

Answer: _____

20 Lymphatic System

OBJECTIVES

1. List three functions of the lymphatic system.
2. Describe the composition and flow of lymph.
3. List the lymphatic organs and lymphatic nodules.
4. Describe the lymph nodes, and state the location of the cervical nodes, axillary nodes, and inguinal nodes.

Part I: Mastering the Basics

MATCHING

Organs of the Lymphatic System

Directions: Match the following terms to the most appropriate definition by writing the correct letter in the space provided. Some terms may be used more than once. See text, pp. 376-380.

A. subclavian veins	F. thymosin
B. thoracic duct	G. thymus
C. right lymphatic duct	H. spleen
D. MALT	I. lymph
E. lymph nodes	

1. _____ A gland located in the upper thorax; it is most active during early life and is concerned with the processing and maturation of T lymphocytes

2. _____ Mucosa-associated lymphatic tissue

3. _____ Lymph from the right arm and the right side of the head and thorax drain into this large duct.

4. _____ Hormone secreted by the thymus gland

5. _____ Acts as a storage reservoir for blood

6. _____ Small, pea-shaped structures that filter lymph as it flows through lymphatic vessels

7. _____ Abdominal organ that resembles a large lymph node; filters blood rather than lymph

8. _____ Most of the lymph of the body drains into this large duct.

9. _____ Contains red pulp and white pulp; it is the largest lymphatic organ in the body

10. _____ The right lymphatic duct and the thoracic duct empty lymph into these blood vessels.

11. _____ Tissue fluid that enters the lymphatic vessels

DRAW IT

Directions: Using this illustration, draw in the following.

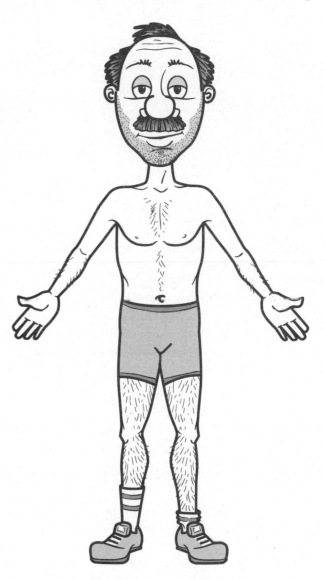

1. The thymosin-secreting gland that involutes during puberty

2. The organ that functions as a giant lymph node; called the *graveyard of the red blood cells*

3. Chain of inguinal lymph nodes

4. Chain of cervical lymph nodes

5. Chain of axillary lymph nodes

6. Place an *X* over the lymph nodes that may be affected with tonsillitis.

7. Place a *Y* over the lymph nodes that are often biopsied/removed during a mastectomy.

8. Place a *Z* over the lymph nodes that may be removed or irradiated in a patient with a pelvic tumor.

9. Shade in the area (light blue) that is drained by the thoracic duct.

10. Shade in the area (yellow) that is drained by the right lymphatic duct.

MATCHING

Lymph Nodes and Tonsils

Directions: Match the following terms to the most appropriate definition by writing the correct letter in the space provided. See text, pp. 378, 379.

A. lingual   D. pharyngeal
B. cervical   E. axillary
C. inguinal   F. palatine

1. _____ Lymph nodes that drain the upper extremities, shoulders, and breast areas

2. _____ Small masses of lymphatic tissue (tonsils) located laterally at the opening of the oral cavity into the pharynx

3. _____ Tonsils located at the back of the tongue

4. _____ Lymph nodes that drain the head and neck area

5. _____ Tonsil located near the opening of the nasal cavity in the upper pharynx; also called an *adenoid*

6. _____ Lymph nodes that drain lymph from the lower extremities and the groin area

Part II: Putting It All Together

MULTIPLE CHOICE

Directions: Choose the correct answer.

1. The lymph nodes, thymus gland, spleen, and tonsils
   a. are confined to the thoracic cavity.
   b. are confined to the ventral cavity.
   c. are lymphatic tissue.
   d. synthesize erythrocytes.

2. Which of the following best describes a tonsil?
   a. a partially encapsulated lymph node
   b. inguinal lymph nodes
   c. white pulp and red pulp
   d. Peyer's patches

3. An adenoid is
   a. a tonsil.
   b. located within the inguinal region.
   c. located within the mediastinum.
   d. an axillary lymph node.

4. What is the name of the fluid from which lymph is made?
   a. aqueous humor
   b. cerebrospinal fluid
   c. cytoplasm
   d. tissue or interstitial fluid

5. If the spleen becomes overactive, it prematurely removes platelets from the circulation, causing thrombocytopenia. As a result, the person is likely to develop
   a. excessive bleeding.
   b. jaundice.
   c. infectious mononucleosis.
   d. lymphedema.

6. On which set of tonsils is a tonsillectomy most often performed?
   a. mucosa-associated lymphatic tissue (MALT)
   b. palatine
   c. pharyngeal
   d. lingual

7. Which of the following is least true of the spleen?
   a. phagocytoses old, worn-out red blood cells
   b. lymphatic organ
   c. a vital organ (cannot live without the spleen)
   d. left upper quadrant (LUQ)

8. A mastectomy with lymph node dissection is likely to cause
   a. lymphedema.
   b. thrombocytopenia and bleeding.
   c. granulocytopenia and infection.
   d. platelet deficiency.

9. Cancer of the breast first metastasizes to the
   a. tonsils.
   b. spleen.
   c. axillary lymph nodes.
   d. thymus gland.

10. The thoracic duct
    a. empties lymph into the left subclavian vein.
    b. receives lymph from the entire right side of the body.
    c. contains blood.
    d. is the main duct within the spleen.

11. Which of the following is most true of a sentinel node biopsy?
    a. diagnostic for the possibility of metastasis of cancer to the breast
    b. assessment tool for the severity of lymphedema
    c. therapeutic technique for shrinking axillary lymph nodes
    d. therapeutic technique whereby radioisotopes are used to destroy cancerous breast tissue

12. Which group is incorrect?
    a. collection of lymph nodes: cervical, axillary, inguinal
    b. tonsils: palatine, pharyngeal, spleen
    c. lymphatic tissue: lymph nodes, tonsils, thymus gland, appendix, MALT
    d. large lymphatic ducts: right lymphatic duct, thoracic duct

13. Which group is incorrect?
    a. collection of lymph nodes: cervical, axillary, inguinal
    b. tonsils: palatine, pharyngeal, lingual
    c. lymphatic tissue: lymph nodes, MALT, thymus gland, spleen
    d. large lymphatic ducts: right lymphatic duct, thoracic duct, venae cavae

CASE STUDY

M.K. is an 18-year-old college freshman. During exam week, he developed a high fever, sore throat, general malaise, productive cough, and small painful lump in his neck. A physician diagnosed him as having the flu and prescribed bed rest, aspirin, and fluids.

1. What is the most likely cause of the painful lump?
   a. A neck muscle was sprained during a bout of coughing.
   b. A lymph node became inflamed as it was fighting the infection.
   c. The painful lump is unrelated to the infection and is probably caused by cancer of the lymph nodes.
   d. The painful lump is a blood-filled cyst.

2. Which of the following is most descriptive of the painful lump?
   a. adenoid
   b. malignant neoplasm
   c. cervical lymph node
   d. thoracic duct

3. What should eventually happen to the painful lump?
   a. It should metastasize.
   b. It should continue to grow and eventually require surgical removal.
   c. As the infection clears up, the lump and pain should disappear.
   d. The surface of the lump should ulcerate, thereby allowing drainage of pus.

Hint: Aching Adenoids

Directions: Perform the following functions on the Sequence of Words that follows. When all the functions have been performed, you are left with a word or words related to the hint. Record your answer in the space provided.

Functions: Remove the following:

1. The three sets of tonsils

2. The large lymphatic organ that resides in the LUQ

3. The two large lymphatic ducts that drain into the subclavian veins

4. Three groups of lymph nodes

5. Lymphatic organ located within the mediastinum; involutes with age

6. Elephantiasis is a form of . . . .

7. The origin of lymph

8. Mucosal-associated lymphatic tissue

9. Hormones secreted by the thymus gland

10. Collection of lymphatic nodules scattered throughout the large intestine

Sequence of Words

L Y M P H E D E M A T H Y M U S G L A N D P E Y
E R S P A T C H E S C E R V I C A L P A L A T I N E
T H Y M O S I N S S P L E E N I N G U I N A L R I G
H T L Y M P H A T I C D U C T I N T E R S T I T I A
L F L U I D L I N G U A L T O N S I L L I T I S M A L
T P H A R Y N G E A L A X I L L A R Y T H O R A C
I C D U C T

Answer: _____

21 Immune System

OBJECTIVES

1. Discuss nonspecific immunity, including:
   - Describe the process of phagocytosis.
   - Explain the causes of the signs of inflammation.
   - Explain the role of fever in fighting infection.
   - Explain the role of interferons, complement proteins, and natural killer cells in the defense of the body.
2. Discuss specific immunity, including:
   - Differentiate between specific and nonspecific immunity.
   - Explain the role of T cells in cell-mediated immunity.
   - Explain the role of B cells in antibody-mediated immunity.
3. Differentiate between genetic immunity and acquired immunity.
4. Describe naturally and artificially acquired active and passive immunity.
5. Describe other immune responses, including:
   - Identify the steps in the development of anaphylaxis.
   - Define *autoimmunity*.

Part I: Mastering the Basics

MATCHING

Nonspecific Immunity

Directions: Match the following terms to the most appropriate definition by writing the correct letter in the space provided. Some terms may be used more than once. See text, pp. 385-389.

A. inflammation
B. natural killer (NK) cells
C. mechanical barriers
D. phagocytes
E. protective proteins
F. chemical barriers
G. fever
H. reflexes
I. temperature

1. _____ Classic symptoms are redness, heat, swelling, and pain.

2. _____ Caused by pyrogens

3. _____ Special type of lymphocyte that acts nonspecifically to kill certain cells

4. _____ Regulated by a hypothalamic thermostat

5. _____ Describes the white blood cells (WBCs), particularly the neutrophils and monocytes, that wander around the body and engage in cellular eating

6. _____ Treated with an antipyretic drug such as aspirin

7. _____ Examples are sneezing, coughing, and vomiting.

8. _____ Examples are lysozyme found in tears and hydrochloric acid in the stomach, saliva, and perspiration.

9. _____ Examples are interferons and complement proteins.

10. _____ Examples are intact skin and mucous membranes.

MATCHING

Specific Immunity

Directions: Match the following terms to the most appropriate definition by writing the correct letter in the space provided. Some terms may be used more than once. See text, pp. 389-392.

A. antigen
B. antigen presentation
C. immunotolerance
D. lymphocytes
E. lymphokines
F. plasma cell
G. cell-mediated immunity
H. autoimmune disease
I. macrophage
J. clone
K. thymus gland
L. antibody-mediated immunity
M. bone marrow
N. agglutination

1. _____ T and B cells

2. _____ A foreign substance, usually a large protein, that provokes a specific immune response

3. _____ The cell-to-cell contact whereby T cells attack antigens directly

4. _____ Also called *humoral immunity*

5. _____ Recognition of self

6. _____ The cell responsible for antigen presentation

7. _____ Attack against self

8. _____ Type of immunity associated with immunoglobulins

9. _____ Site where T cells mature and differentiate

10. _____ Site where B cells mature and differentiate

169

11. _____ A group of cells formed from the same parent cell

12. _____ Reacts with an antibody to cause clumping

13. _____ An allergen

14. _____ Chemicals secreted by activated T cells that enhance phagocytic activity

15. _____ A subgroup of the B-cell clone that produces antibodies

16. _____ Ability of the macrophage to push an antigen to its surface

READ THE DIAGRAM

Immunoglobulin E (IgE) Hypersensitivity Reaction

Directions: Referring to the illustration, fill in the blanks with the numbers. See text, pp. 395, 396.

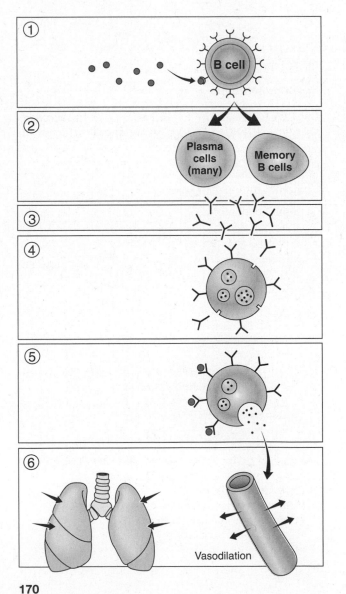

1. _____ Illustrates the consequences of histamine release

2. _____ Antigens bind to the receptors on the B cell.

3. _____ Antigens bind to IgE antibodies on the mast cells.

4. _____ Histamine is released from mast cell

5. _____ Antibodies are secreted.

6. _____ The intact mast cell with its IgE antibodies

7. _____ B-cell clone that secretes antibodies

8. _____ Illustrates an anaphylaxis-induced dyspnea (difficulty in breathing)

9. _____ A circulating lymphocyte

10. _____ Illustrates an anaphylaxis-induced hypotension

11. _____ Illustrates why the mast cell bursts

MATCHING

Types of Immunity

Directions: Match the following terms to the most appropriate definition by writing the correct letter in the space provided. Some terms may be used more than once. See text, pp. 390-392.

A. titer
B. immunization
C. genetic immunity
D. secondary response
E. primary response
F. naturally acquired active immunity
G. naturally acquired passive immunity
H. artificially acquired active immunity
I. artificially acquired passive immunity

1. _____ Inborn, inherited, or species immunity

2. _____ A type of immunity achieved by getting an injection of gamma globulin

3. _____ An infant receives antibodies (IgA) from her or his mother while breastfeeding.

4. _____ Level of antibodies in the blood

5. _____ Immunization achieved by the injection of attenuated measles virus

6. _____ An infant received antibodies (IgG) from his or her mother while in utero.

7. _____ Immunization achieved by injection of a toxoid

8. _____ The type of immunity stimulated by a vaccine

9. _____ The type of natural immunity that mandates that you will not get the disease if exposed to the virus as an adult because you had the disease as a child

10. _____ Initial response of the plasma cells to an antigen

11. _____ Related to a booster shot

12. _____ The type of immunity that mandates that you will not get Dutch elm disease from your tree

13. _____ Also called *vaccination*

14. _____ Antitoxins, antivenoms

SIMILARS AND DISSIMILARS

Directions: Circle the word in each group that is least similar to the others. Indicate the similarity of the three words on the line below each question.

1. cytotoxic cells   helper cells   memory cells   granulocytes

_____

2. B cells   antibody-mediated immunity (AMI)   hemoglobin   plasma cells

_____

3. T cells   cell-mediated immunity (CMI)   nonspecific immunity   B and T lymphocytes

_____

4. phagocytosis   diapedesis

   anaphylaxis   chemotaxis

_____

5. phagocytosis   T and B cells   fever   inflammation

_____

6. specific immunity   fever   pyrogen   interferons

_____

7. redness   heat   jaundice   swelling

_____

8. complement protein   interferons   protective proteins   pyrogen

_____

9. T cells   nonspecific immunity   B cells   plasma cells

_____

10. IgG   antibodies   neutrophils   IgA

_____

11. toxoid   passive immunity   vaccine   getting measles

_____

12. histamine   IgE   anaphylaxis   leukocytosis antibodies

_____

13. immune globulin   antitoxin   antivenom   vaccine

_____

14. CMI   AMI   B and T lymphocytes   inflammation

_____

Part II: Putting It All Together

MULTIPLE CHOICE

Directions: Choose the correct answer.

1. When a pathogen causes redness, heat, swelling, and pain, the process is called
   a. chemotaxis.
   b. diapedesis.
   c. an anaphylactic reaction.
   d. an infection.

2. Which of the following is most related to pus?
   a. hyperbilirubinemia and jaundice
   b. phagocytosis and infection
   c. autoimmune disease
   d. thrombocytopenia and bleeding

3. Leukocytes "go" to the site of infection. Which of the following is descriptive of this activity?
   a. fever and flushing
   b. histamine and urticaria
   c. diapedesis and chemotaxis
   d. mitosis and neoplasia

4. Which of the following is most descriptive of a Kupffer cell?
   a. It is a helper T cell.
   b. It is a "fixed" macrophage located within the liver.
   c. It is a memory cell.
   d. It secretes antibodies.

5. Which of the following speeds up the recognition of an antigen (allergen) when it is introduced for a second and third time?
   a. suppressor T cell
   b. neutrophil
   c. phagocyte
   d. memory cell

6. What is the term that refers to the ability of our bodies to protect against pathogens and other foreign agents?
   a. diapedesis
   b. chemotaxis
   c. complement fixation
   d. immunity

7. Which of the following includes mechanical barriers, chemical barriers, and certain reflexes?
   a. specific defense mechanisms
   b. third line of defense
   c. nonspecific defense mechanisms
   d. interferons and complement proteins

8. The "big-eating" fixed phagocytes located in the liver, spleen, and lymph nodes
   a. are called NK cells.
   b. are B cells.
   c. are macrophages.
   d. develop in response to a specific antigen.

9. Which of the following is an immune system stimulator?
   a. use of anticancer drugs
   b. use of steroids
   c. aging
   d. administration of a booster shot

10. Fever-producing substances (pyrogens)
    a. are released by phagocytes as they perform their tasks.
    b. cause the release of histamine.
    c. are called *interferons*.
    d. are immunoglobulins.

11. Interferons are
    a. proteins secreted by a cell infected by a virus so as to prevent further viral replication.
    b. complement proteins.
    c. called *IgE antibodies*.
    d. secreted by T cells.

12. Your own cells and secretions are recognized by your body as nonantigenic. What is this self-recognition called?
    a. immunity
    b. immunocompetence
    c. immunotolerance
    d. autoimmunity

13. Antibody-mediated immunity
    a. is achieved by B cells.
    b. operates by cell-to-cell combat.
    c. always results in anaphylaxis.
    d. is achieved by NK cells.

14. Which of the following is true of the T lymphocyte?
    a. Engages in antibody-mediated immunity
    b. Activated in response to a specific antigen
    c. Its clone includes plasma cells and memory T cells.
    d. Is myelogenous, granulocytic, phagocytic, and anti-inflammatory

15. Which of the following is true of the B lymphocyte?
    a. Activated nonspecifically in response to all antigens
    b. Engages in cell-mediated conflict
    c. Called a Kupffer cell when "fixed" within the hepatic sinusoids
    d. Its clone includes antibody-secreting plasma cells.

16. To which are B- and T-lymphocyte activity most related?
    a. endocytosis and phagocytosis
    b. antigens and antibodies
    c. interferons and complement proteins
    d. AMI and CMI

17. Which of the following is least related to anaphylaxis?
    a. is associated with IgE antibodies
    b. refers to the cluster of symptoms of an acute allergic reaction
    c. is characterized by life-threatening respiratory responses
    d. is also called immunotolerance

18. Which of the following causes the release of histamine and leukotrienes?
    a. binding of the allergen (antigen) to the IgE antibodies on the mast cells
    b. ingestion of a pathogen by a macrophage
    c. Puncturing of the bacterial cell wall by complement protein
    d. binding of the allergen to the antigen

CASE STUDY

While romping through the fields one day, Aunty Bea was attacked and stung by an angry swarm of bees. Knowing that she was allergic to bee venom, her friends rushed her to the nearest emergency department. By the time she reached the emergency department, Bea's lips and tongue were swollen, and she was unable to speak. Her breathing was labored, and she was becoming cyanotic. She was immediately given an injection of epinephrine and steroids, intubated, and placed on oxygen.

1. What process is involved in the allergic response?
   a. the release of lysozyme
   b. the release of histamine in response to IgE antibodies
   c. only the nonspecific defense mechanisms
   d. only the CMI response

2. Why was Bea dyspneic (having difficulty breathing)?
   a. The bee venom caused widespread vasodilation.
   b. The bee venom paralyzed olfactory receptors in the nasal passages.
   c. The swelling of the respiratory passages blocked the movement of air in and out of the lungs.
   d. She was too anxious to breathe normally.

3. The oxygen was administered to
   a. block the release of additional histamine.
   b. activate the steroid.
   c. neutralize the bee venom.
   d. correct the hypoxemia and cyanosis.

4. Why was epinephrine administered?
   a. It was given to dilate the respiratory passages.
   b. It was given to neutralize the bee venom.
   c. It was given to lower the blood pressure.
   d. It was given to help the liver degrade the histamine.

5. Which of the following is true of histamine?
   a. It is released from plasma cells.
   b. It causes constriction of the respiratory passages and decreases the flow of air.
   c. It causes vasoconstriction and an increase in blood pressure.
   d. It should be administered intravenously to combat the allergic response to the bee venom.

PUZZLE

Hint: Inflammation: Antibody-generating

Directions: Perform the following functions on the Sequence of Words that follows. When all the functions have been performed, you are left with a word or words related to the hint. Record your answer in the space provided.

Functions: Remove the following:

1. Immunoglobulins (five)

2. Cells (two) that comprise the specific immune response

3. Protective proteins (two)

4. Convey active immunity (two)

5. Conveys passive immunity (one)

6. Antibody-secreting B cells

7. The type of lymphocyte that is necessary for both B- and T-lymphocyte activity

8. The immediate onset and most severe allergic reaction

9. Fever producing

10. Pus producing

11. Inflammation caused by a pathogen

12. Means hives

13. Four classic signs of inflammation

14. Processes (two) that describe the movement and attraction of the WBC to the site of infection

Sequence of Words

P Y O G E N I C I D R E D N E S S I N F E C T I O N
H E A T P Y R O G E N I C S W E L L I N G G A M M
A G L O B U L I N I N T E R F E R O N S T L Y M P
H O C Y T E S C O M P L E M E N T P R O T E I N S
D I A P E D E S I S U R T I C A R I A I G T O X O I D
I M V A C C I N E Ig A P A I N C H E M O T A X I S
Ig E B L Y M P H O C Y T E S P L A S M A C E L L
S H E L P E R A N T I G E N A N A P H Y L A X I S

Answer: _____

BODY TOON

Hint: Avian-Viral Event

Answer: bird flew (bird flu)

22 Respiratory System

Answer Key: Textbook page references are provided as a guide for answering these questions. A complete answer key was provided for your instructor.

OBJECTIVES

1. Describe the structure and functions of the organs of the respiratory system, and trace the movement of air from the nostrils to the alveoli.
2. Describe why lungs collapse or expand and the role of pulmonary surfactants.
3. Discuss the three steps in respiration, including:
   - Describe the relationship of Boyle's law to ventilation.
   - Explain how respiratory muscles affect thoracic volume.
   - List three conditions that make the alveoli well suited for the exchange of oxygen and carbon dioxide.
4. List lung volumes and capacities.
5. Discuss the voluntary and involuntary control of breathing, including:
   - Explain the neural and chemical control of respiration.
   - Describe common variations and abnormalities of breathing.

Part I: Mastering the Basics

MATCHING

Structures of the Respiratory System

Directions: Match the following terms to the most appropriate definition by writing the correct letter in the space provided. Some terms may be used more than once. See text, pp. 403-411.

A. larynx
B. trachea
C. nose
D. pharynx
E. paranasal sinuses
F. lungs
G. alveoli
H. bronchioles
I. bronchi
J. glottis
K. esophagus
L. carina

1. _____ Called the *windpipe*, it is a strong cartilaginous tube that conducts air to and from the lungs.

2. _____ Respiratory structure that communicates with the middle ear by the eustachian tube

3. _____ The Adam's apple or thyroid cartilage is most associated with this structure.

4. _____ The epiglottis directs food and water from the respiratory passages into this structure.

5. _____ Called the *throat*

6. _____ Called the *voice box* because it contains the vocal cords

7. _____ The point at which the trachea bifurcates (splits); the area is extremely sensitive and elicits coughing when stimulated (as in suctioning with a catheter)

8. _____ Located between the larynx and the bronchi and anterior to the esophagus

9. _____ Composed of three parts—naso-, oro-, and laryngo-

10. _____ Large tube that splits into bronchi

11. _____ These small structures located within the bronchial tree are composed primarily of smooth muscle.

12. _____ The exchange of the respiratory gases between the air and blood occurs here.

13. _____ Large, soft, cone-shaped organs that contain the respiratory passages and pulmonary capillaries; they fill most of the thoracic cavity

14. _____ Because of smooth muscle, this structure can contract and relax, thereby causing constriction and dilation.

15. _____ The olfactory receptors are located within this structure.

16. _____ Mucus drains into the nasal cavities from these structures located in the head.

17. _____ The trachea splits into the right and left.

18. _____ Small respiratory passages that deliver oxygen to the alveoli

19. _____ Structures partially encircled by the pulmonary capillaries

20. _____ Called the *resistance vessels*

21. _____ Structures that contain surfactants

22. _____ The space between the vocal cords

Structures of the Respiratory Tract

Directions: Referring to the illustration, fill in the blanks on the next page with the correct numbers. See text, pp. 403-411.

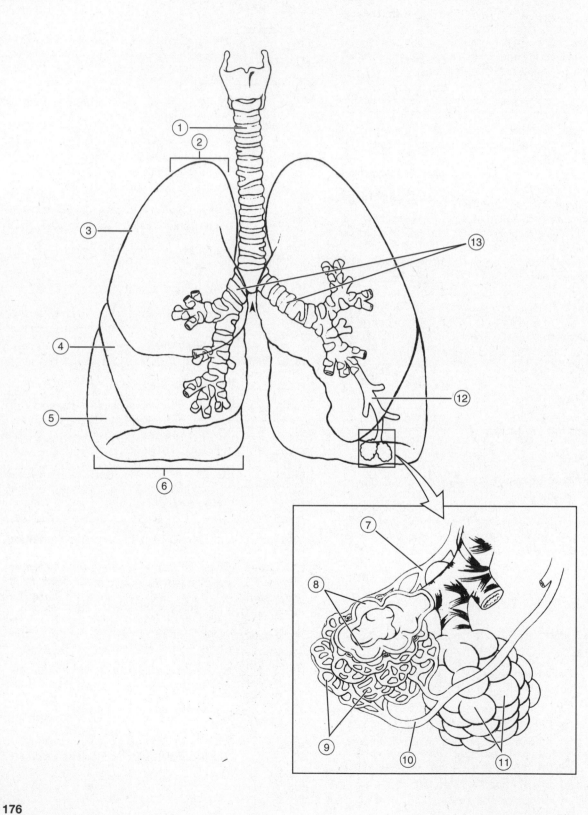

1. _____ Structure that delivers air to the bronchi

2. _____ Pulmonary capillaries partially surround these grapelike structures.

3. _____ Structure that delivers oxygen to the alveoli

4. _____ Lung structure concerned with the exchange of O_2 and CO_2

5. _____ Apex of the lung

6. _____ Inferior lobe of the right lung

7. _____ Trachea branches into these large structures

8. _____ Base of the lung

9. _____ Middle lobe of the right lung

10. _____ Grapelike structures that contain surfactants

11. _____ Windpipe; kept open by rings of cartilage

12. _____ Pulmonary capillaries

13. _____ Superior lobe of the right lung

14. _____ Large tubes that deliver air to the bronchioles

15. _____ Blood vessel branch that delivers oxygenated blood to the left atrium

16. _____ Blood vessel branch that delivers unoxygenated blood to the alveoli

17. _____ Blood vessels that surround the alveoli; participate in gas exchange

18. _____ Blood vessel branch that comes from the right ventricle of the heart

COLORING

Directions: Color the appropriate areas on the illustration on the previous page.

1. Color the right lung *pink*.

2. Color the left lung *yellow*.

3. Color the conducting airways *light brown*.

4. Color the alveoli *green*.

5. Color the bronchioles *light green*.

6. Color the cartilaginous rings along the bronchi *black*.

7. Color the pulmonary artery branch *blue*.

8. Color the pulmonary vein branch *red*.

9. Color the pulmonary capillaries *purple*.

ORDERING

Air Flow During Inhalation

Directions: Place in the correct order the structures through which air flows during inhalation. The end point (alveoli) has been inserted.

Structures

Bronchioles

Pharynx

Trachea

Alveoli

Nasal cavities

Bronchi

Larynx

_____

_____

_____

_____

_____

_____

Alveoli _____

MATCHING

Thoracic Cavity, Pleural Cavity, and Mediastinum

Directions: Match the following terms to the most appropriate definition by writing the correct letter in the space provided. Some terms may be used more than once. See text, pp. 411-418.

A. intrapleural space   F. thoracic cavity
B. visceral pleura   G. intercostal muscles
C. diaphragm   H. acetylcholine
D. phrenic   I. mediastinum
E. nicotinic

1. _____ Membrane on the outer surface of each lung

2. _____ Space between the visceral and parietal pleural membranes; also called a *potential space*

177

3. _____ Muscle that separates the thoracic cavity from the abdominal cavity

4. _____ Area between the two lungs; contains other thoracic structures, such as the heart, great vessels, and trachea

5. _____ Dome-shaped muscle that is the chief muscle of inhalation

6. _____ For the lungs to remain expanded, the pressure must be negative in this area.

7. _____ Contains the pleural cavities, pericardial cavity, and mediastinum

8. _____ Skeletal muscles between the ribs; move the rib cage up and out during inhalation

9. _____ A pneumothorax occurs when air enters this area.

10. _____ Neurotransmitter at the neuromuscular junction (diaphragm and phrenic nerve)

11. _____ Motor nerve that stimulates the diaphragm

12. _____ Receptor within the neuromuscular junction (phrenic nerve and diaphragm)

MATCHING

Pulmonary Volumes and Capacities

Directions: Match the following terms to the most appropriate definition by writing the correct letter in the space provided. Some terms may be used more than once. See text, pp. 419-422.

A. residual volume
B. tidal volume
C. spirometer
D. vital capacity
E. dead air space
F. peak expiratory flow rate
G. expiratory reserve volume
H. inspiratory reserve volume
I. forced expiratory volume (FEV)

1. _____ A combination of tidal volume, inspiratory reserve volume, and expiratory reserve volume, about 4600 mL

2. _____ The amount of air that remains in the lungs after the exhalation of the expiratory reserve volume, about 1200 mL; this air cannot be exhaled

3. _____ An instrument that measures pulmonary volumes

4. _____ The amount of air moved into or out of the lungs with each breath; the average is 500 mL at rest

5. _____ The amount of air you can inhale after a normal inhalation, about 3000 mL

6. _____ A pulmonary capacity that is the maximal amount of air exhaled after maximal inhalation

7. _____ The additional volume of air that you can exhale after a normal exhalation

8. _____ The volume of air that you move during normal quiet breathing

9. _____ The fraction of the vital capacity exhaled in a specific number of seconds (i.e., FEV_1)

10. _____ Defined as the maximal rate of air flow during expiration

11. _____ The air that remains in the conducting spaces of the respiratory tract (trachea, bronchi, and bronchioles); it is not available for exchange (about 150 mL)

12. _____ The following are instructions for its use: "Take the deepest breath possible. Exhale all the air you possibly can into this tube."

ORDERING

Ventilation (Inhalation and Exhalation)

Directions: Place in order the events that occur during one ventilatory cycle (inhalation and exhalation) by writing the correct phrase in the space provided. Two of the steps are given. See text, pp. 416-420.

1. The inspiratory neurons in the brain fire action potentials (nerve impulses).

2. Pressure within the lungs (intrapulmonic pressure) decreases.

3. The nerve impulses travel along the phrenic and intercostal nerves to the diaphragm and the intercostal muscles.

4. The diaphragm and the intercostal muscles relax.

5. Air leaves the lungs (air is exhaled).

6. Thoracic volume decreases.

7. The pressure within the lungs (intrapulmonic pressure) increases.

8. The diaphragm and the intercostal muscles contract, thereby enlarging the thoracic cavity.

9. Air moves into the lungs (air is inhaled).

10. The phrenic and intercostal nerves stop firing.

1. The inspiratory neurons in the brain fire action potentials (nerve impulses).

2. _____

3. _____

4. _____

5. Air moves into the lungs (air is inhaled).

6. _____

7. _____

8. _____

9. _____

10. _____

SIMILARS AND DISSIMILARS

Directions: Circle the word in each group that is least similar to the others. Indicate the similarity of the three words on the line below each question.

1. trachea   pharynx   bronchus   alveolus

_____

2. voice box   phrenic   larynx   vocal cords

_____

3. lobes   nares   apex   base

_____

4. parietal pleura   serous membrane   pleurae   mucous membrane

_____

5. polar water molecule   surfactants   surface tension   bronchus

_____

6. inhalation   ventilation   expiration   anatomic dead space

_____

7. inspiration   inhalation   epiglottis   ventilation

_____

8. Boyle's law   pressure   volume   phrenic

9. intercostal   phrenic   diaphragm   trachea
   muscles

_____

10. oxyhemoglobin   thrombocyte

    carbaminohemoglobin   RBCs

_____

11. tidal volume   vital capacity

    inspiratory reserve volume   intrapleural space

_____

12. medulla oblongata   pneumotaxic center

    apneustic center   diaphragm

_____

13. hypoxemia   jaundice   cyanosis   apnea

_____

14. carina   dyspnea   hypoxemia   orthopnea

_____

Part II: Putting It All Together

MULTIPLE CHOICE

Directions: Choose the correct answer.

1. Which of the following is least descriptive of the alveoli?
   a. grapelike sacs located very close to the pulmonary capillaries
   b. primarily concerned with gas exchange
   c. located within the lungs
   d. contain smooth muscle

2. Bronchi, bronchioles, and alveoli
   a. have rings of cartilage surrounding them.
   b. are sites of gas exchange.
   c. are composed entirely of smooth muscle.
   d. are all located within the lungs.

3. The epiglottis
   a. prevents food and water from entering the respiratory passages.
   b. secretes surfactants.
   c. is considered a true vocal cord.
   d. is concerned with exchange of the respiratory gases.

4. Which of the following refers to the amount of air maximally exhaled after maximal inhalation?
   a. tidal volume
   b. total lung capacity
   c. expiratory reserve volume
   d. vital capacity

5. The trachea does not collapse because it is
   a. located within the lungs.
   b. composed of smooth muscle.
   c. surrounded by ribs.
   d. composed of tough cartilaginous rings.

6. In the absence of surfactants,
   a. it is difficult to open the alveoli.
   b. bronchioles relax.
   c. the alveoli fill with water.
   d. air enters the intrapleural space.

7. What is the effect of contraction of the diaphragm and intercostal muscles?
   a. increases the volume of the thoracic cavity
   b. forces air out of the lungs
   c. closes the epiglottis
   d. stimulates the secretion of pulmonary surfactants

8. The phrenic nerve
   a. causes relaxation of the bronchioles.
   b. supplies the epiglottis so that food is diverted to the esophagus.
   c. stimulates the diaphragm to contract.
   d. exits from the spinal cord at the L3 level.

9. Which transport mechanism causes the respiratory gases to move across the alveolar-pulmonary capillary membrane?
   a. filtration
   b. osmosis
   c. diffusion
   d. active transport pump

10. Why does a stab wound to the chest cause the lung to collapse?
    a. Infection develops within the intrapleural space.
    b. The intrapleural pressure increases.
    c. Pressure within the lung increases.
    d. Bleeding causes hypotension and septic shock.

11. What does Boyle's law state?
    a. When volume increases, pressure increases.
    b. When volume decreases, pressure decreases.
    c. When volume increases, pressure decreases.
    d. A change in volume has no effect on pressure.

12. How is most oxygen transported in the blood?
    a. by hemoglobin
    b. as bicarbonate
    c. by pulmonary surfactants
    d. dissolved in plasma

13. How is most carbon dioxide transported in the blood?
    a. by hemoglobin
    b. as bicarbonate
    c. by pulmonary surfactants
    d. dissolved in plasma

14. The medullary respiratory control center
    a. is located in the cerebrum.
    b. is sensitive to the depressant effects of opioids (narcotics).
    c. can increase respiratory rate but cannot decrease respiratory rate.
    d. can decrease respiratory rate but cannot increase respiratory rate.

15. An increase in blood _____ is most likely to increase the rate of breathing.
    a. volume
    b. CO_2
    c. bilirubin
    d. calcium

16. Orthopnea refers to
    a. a lung infection like pneumonia.
    b. a collection of excess fluid within the intrapleural space.
    c. a bluish color of the skin induced by hypoxemia.
    d. difficulty in breathing that is relieved by sitting upright.

17. Bradypnea and tachypnea
    a. refer to breathing rates.
    b. are respiratory changes caused by exercise.
    c. are upper respiratory infections.
    d. are relieved by sitting upright.

18. Which of the following are most apt to cause hypoxemia?
    a. Kussmaul respirations, hyperventilation, apnea
    b. hypoventilation, apnea, medullary opioid depression
    c. tachypnea, Kussmaul respirations, hypoventilation
    d. bradypnea, hyperventilation, cyanosis

19. Activation of the respiratory beta$_2$ adrenergic receptors
    a. causes bronchoconstriction.
    b. relaxes the diaphragm and increases thoracic volume.
    c. dilates the bronchioles.
    d. increases the secretion of the pulmonary surfactants.

20. Which of the following is most descriptive of ventilation?
    a. inhalation, inspiration
    b. exhalation, expiration
    c. inhalation, exhalation
    d. diaphragm, intercostal nerve

21. Which of the following is a correct air flow pattern?
    a. bronchus → trachea → alveoli
    b. trachea → bronchi → bronchioles
    c. bronchioles → glottis → alveoli
    d. larynx → alveoli → carina

22. "Clubbing" of the fingers is due to
    a. CO$_2$ retention and acidosis.
    b. exercise.
    c. chronic hypoxemia.
    d. acute sinus infection.

23. A patient is described as dyspneic. He is
    a. coughing up blood (hemoptysis).
    b. flushed.
    c. not breathing.
    d. having difficulty in breathing.

24. If a person is apneic, she is
    a. hyperventilating.
    b. experiencing Kussmaul respirations.
    c. not breathing.
    d. orthopneic.

25. What is the consequence of a severed spinal cord at the level of C2?
    a. quadriplegia with no respiratory impairment
    b. hemiplegia with no respiratory involvement
    c. quadriplegia and respiratory paralysis
    d. paraplegia with no respiratory involvement

26. Which of the following statements is true about the aging respiratory system?
    a. By the age of 70 years, vital capacity has decreased about 33%.
    b. Aging is associated with a decrease in vital capacity and with an increase in the total number of alveoli.
    c. By the age of 70 years, all the alveoli have collapsed, and gas exchange is occurring across the walls of the bronchioles.
    d. Because of the decrease in vital capacity, a person older than 70 years should not exercise.

CASE STUDY

T.K. went to the doctor's office complaining of extreme fatigue, shortness of breath, fever, and a persistent cough of 3 weeks' duration. An x-ray revealed left lower lobar pneumonia with areas of alveolar collapse (atelectasis). He was given a prescription for an antibiotic.

1. What is the effect of left lower lobar pneumonia and atelectasis on breathing?
    a. There are fewer alveoli available for gas exchange.
    b. Rapid shallow breathing can easily compensate for the collapsed alveoli; no further treatment is required.
    c. The alveoli in the left lower lobe can function fairly well if they fill with blood from the pulmonary capillaries.
    d. There will be no effect on breathing and P$_{O_2}$.

2. Which of the following is the underlying cause of the fever?
    a. atelectasis
    b. infection
    c. cyanosis
    d. hypoxemia

3. Coughing
    a. is the body's attempt to clear the respiratory passages.
    b. should be completely suppressed to rest the lungs.
    c. should always be suppressed because it could spread the infection to the unaffected lung.
    d. should always be suppressed because it impairs the effectiveness of the antibiotics.

Hint: Sneeze and Wheeze

Directions: Perform the following functions on the Sequence of Words that follows. When all the functions have been performed, you are left with a word or words related to the hint. Record your answer in the space provided.

Functions: Remove the following:

1. The structures that are called the *throat, voice box*, and *windpipe*

2. The terms that describe the breathing in and breathing out phases of ventilation (four)

3. The respiratory gases (two)

4. Serous membranes (two) located within the thoracic cavity

5. Maximal exhalation after maximal inhalation

6. Grapelike respiratory structure concerned with the exchange of O_2 and CO_2

7. When volume increases, pressure decreases.

8. Primary muscle of inhalation and its motor nerve

9. Amount of air moved during normal quiet breathing

10. A color consequence of hypoxemia

11. Two pontine control centers for breathing

12. Two ways that O_2 and CO_2 are transported through the blood

13. Secreted by the alveolar cells to decrease surface tension

14. Four abnormal respiratory sounds

Sequence of Words

ALVEOLUSCYANOSISBRONCHOCONS
TRICTIONINSPIRATIONOXYHEMOGLO
BINDIAPHRAGMVITALCAPACITYRALES
CARBONDIOXIDERHONCHIVISCERALP
LEURAPNEUMOTAXICCENTERTRACH
EAEXHALATIONSURFACTANTSPHRENI
CBOYLE'SLAWHCOALLERGYOXYGENA
PNEUSTICCENTERPHARYNXLARYNXI
NHALATIONTIDALVOLUMESTRIDORPA
RIETALPLEURAHISTAMINEWHEEZINGE
XPIRATION

Answer: _____, _____, _____

BODY TOON

Hint: What Does This View Have to Do with a Runny Nose?

Answer: both experience rhino-rear (rhinorrhea)

23 Digestive System

OBJECTIVES

1. Discuss the basic anatomy and physiology of the digestive system, including:
   - List four functions of the digestive system.
   - Explain the processes of digestion and absorption.
   - Describe the four layers, nerves, and membranes of the digestive tract.
2. Describe the structure and functions of the organs and accessory organs of the digestive tract.
3. Explain the physiology of digestion and absorption, including:
   - Describe the effects of amylases, proteases, and lipases.
   - Describe the role of bile in the digestion of fats.
   - Describe the effects of the hormones gastrin, cholecystokinin, and secretin.
4. Discuss nutrition concepts, including:
   - Describe five categories of nutrients.
   - Discuss the importance of a balanced diet.

Part I: Mastering the Basics

MATCHING

Terms Related to the Digestive System

Directions: Match the following terms to the most appropriate definition by writing the correct letter in the space provided. Some terms may be used more than once. See text, pp. 432, 433.

A. alimentary canal
B. digestion
C. peristalsis
D. chemical digestion
E. bolus
F. absorption
G. mechanical digestion

1. _____ Process whereby the end products of digestion move across the walls of the alimentary canal into the blood

2. _____ The hollow tube that extends from the mouth to the anus

3. _____ Process whereby food is broken down into simpler substances that can be absorbed

4. _____ Type of digestion that physically breaks food into smaller pieces

5. _____ A rhythmic contraction of the muscles of the digestive tract that moves food forward toward the anus

6. _____ Ball-like mass of food that is mixed with saliva in the mouth; swallowed and turned into chyme in the stomach

7. _____ Type of digestion accomplished by the digestive enzymes

8. _____ Also called the *digestive tract* or the *gastrointestinal (GI) tract*

9. _____ Type of digestion that includes the chewing and mashing of food

MATCHING

Teeth and Related Structures

Directions: Match the following terms to the most appropriate definition by writing the correct letter in the space provided. See text, pp. 434, 435.

A. deciduous
B. permanent
C. crown
D. gingiva
E. pulp
F. dentin
G. root
H. root canal
I. cementum
J. enamel

1. _____ Term for the 20 baby (milk) teeth

2. _____ Part of the tooth embedded within the jawbone

3. _____ Part of the tooth above the level of the gum and covered by enamel

4. _____ Part of the tooth that anchors the root to the periodontal membrane; holds the tooth in place

5. _____ Part of the tooth containing the nerves, blood vessels, and connective tissue; supplies the tooth with sensation and nutrients

6. _____ Term for the gum

7. _____ Hard brittle covering of the crown of the tooth

8. _____ Bonelike material that makes up the bulk of the tooth

9. _____ These 32 teeth replace the baby teeth.

10. _____ The extension of the pulp cavity into the root

MATCHING

Digestive Enzymes, Hormones, and Other Secretions

Directions: Match the following terms to the most appropriate definition by writing the correct letter in the space provided. Some terms may be used more than once. See text, pp. 438-443, 448-454.

A. amylase(s)
B. lipases
C. bile
D. chyme
E. intrinsic factor
F. cholecystokinin (CCK)
G. hydrochloric acid (HCl)
H. chyle
I. secretin
J. proteases
K. gastric juice
L. saliva
M. disaccharidases

1. _____ Secretion of the salivary glands; contains water, electrolytes, mucus, and ptyalin

2. _____ Partially digested food that is mashed into a pastelike consistency in the stomach

3. _____ A deficiency of this substance leads to pernicious anemia.

4. _____ A substance that is secreted by the parietal cells of the stomach; lowers the gastric pH

5. _____ A substance produced by the liver and stored in the gallbladder

6. _____ Enzyme found in saliva; also called *ptyalin*

7. _____ Classification of pepsin, trypsin, chymotrypsin, and enterokinase

8. _____ Sucrase, maltase, lactase

9. _____ Milky white lymph caused by fat absorption

10. _____ Enzymes that digest proteins into small peptides and amino acids

11. _____ Enzymes that digest fats into fatty acids and glycerol

12. _____ A substance secreted by the parietal cells of the stomach; necessary for the absorption of vitamin B_{12}

13. _____ A hormone secreted by the walls of the duodenum; stimulates the pancreas to secrete a bicarbonate-rich substance

14. _____ Enzymes that digest carbohydrates to disaccharides

15. _____ A hormone secreted by the walls of the duodenum; stimulates the pancreas to secrete digestive enzymes and slows gastric emptying

16. _____ Secretion of the glands of the stomach; contains HCl, intrinsic factor, water, electrolytes, and digestive enzymes

17. _____ A hormone secreted by the walls of the duodenum; causes the gallbladder to contract and eject bile into the common bile duct

18. _____ An emulsifying agent

Directions: Fill in the blanks with the numbers on the diagram that correspond to the description provided. See text, pp. 434-435.

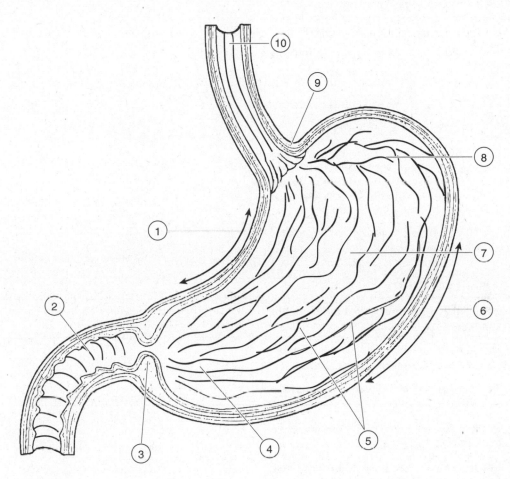

1. _____ The folds of the stomach wall that allow it to expand after a meal

2. _____ The lesser curvature

3. _____ The part of the small intestine that receives chyme from the stomach

4. _____ The domelike part of the stomach

5. _____ The body of the stomach

6. _____ The sphincter most likely to be associated with pyrosis or heartburn

7. _____ The greater curvature

8. _____ The most distal portion of the stomach, the end of which is a sphincter

9. _____ The sphincter that allows food to enter the stomach

10. _____ The part of the stomach that is likely to slide into the chest through a hiatal hernia

11. _____ The sphincter most associated with gastro-esophageal reflux disease (GERD)

12. _____ Reflux of gastric contents enters the base of this "food tube," causing a burning discomfort

13. _____ A narrowing or stenosis of this distal structure in infancy causes projectile vomiting

14. _____ Sometimes called *cardiac sphincter* because of its proximity to the heart

MATCHING

Parts of the Digestive Tract

Directions: Match the following terms to the most appropriate definition by writing the correct letter in the space provided. Some terms may be used more than once. See text, pp. 433-449.

A. stomach
B. small intestine
C. pancreas
D. liver
E. gallbladder
F. buccal cavity
G. salivary glands
H. esophagus
I. epiglottis
J. large intestine

1. _____ Organ that produces bile and secretes it into the hepatic bile ducts

2. _____ Structure that directs food and water away from the larynx into the esophagus

3. _____ The "food tube"; carries food from the pharynx to the stomach

4. _____ The common bile duct empties bile into this structure.

5. _____ Organ that stores many of the fat-soluble vitamins

6. _____ Pear-shaped sac that attaches to the underside of the liver; concentrates and stores bile

7. _____ The parietal cells of this organ secrete HCl and intrinsic factor.

8. _____ Organ that is connected to the common bile duct by the cystic duct

9. _____ Organ that receives blood from the portal vein

10. _____ Parotid, sublingual, and submandibular

11. _____ Organ that is divided into the fundus, body, and pylorus

12. _____ Organ that responds to CCK by ejecting bile

13. _____ Most digestion and absorption occur within this structure.

14. _____ The primary function of this organ is to deliver chyme to the duodenum at the proper rate.

15. _____ Home of the phagocytic Kupffer cells

16. _____ The walls of this organ are thrown into folds called *rugae*.

17. _____ Divisions include the cecum, colon, rectum, and anal canal.

18. _____ Organ that responds to CCK by secreting amylases, proteases (including trypsin), and lipases

19. _____ Organ that secretes the most potent digestive enzymes

20. _____ An inflammation of one of these structures is called *mumps*.

21. _____ Organ that contains mucous cells, parietal cells, and chief cells

22. _____ Organ that responds to secretin by the secretion of bicarbonate, an alkaline secretion

23. _____ The inner lining of this structure is characterized by villi and microvilli; brush border cells

24. _____ Divisions include the duodenum, jejunum, and ileum.

25. _____ Relaxation of the sphincter of Oddi allows bile and pancreatic enzymes to enter this structure.

26. _____ Landmarks of this organ include the greater curvature and lesser curvature.

27. _____ This structure has bends or curves called the *hepatic flexure* and the *splenic flexure*.

28. _____ The walls of this structure secrete CCK.

29. _____ The term *hepatic* refers to this organ.

30. _____ A colostomy is a surgical procedure performed on this structure.

31. _____ The appendix is attached to this structure.

32. _____ The walls of this structure secrete the disaccharidases.

33. _____ Organ that synthesizes most of the plasma proteins, including the clotting factors

34. _____ Part of the mouth

35. _____ The pylorus connects the stomach with this structure.

36. _____ Organ that converts excess glucose into glycogen for storage and use in the regulation of blood sugar

37. _____ Flatus, feces, and defecation are most related to this structure.

38. _____ The chief organ of drug detoxification

39. _____ Organ that secretes insulin, glucagon, trypsin, and a bicarbonate-rich secretion

40. _____ Peristalsis is pendulum-like (sways back and forth) in this organ.

41. _____ Organ that contains the enzymes for urea synthesis

BODY TOON

Hint: Two Punctuation Marks

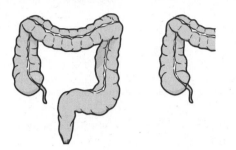

CONNECTIONS

Directions: Indicate the structure that forms the connecting link in the following questions. Some words can be used more than once

A. ascending colon
B. ileum
C. stomach
D. jejunum
E. sigmoid colon
F. hepatic ducts
G. duodenum
H. transverse colon
I. cystic duct
J. descending colon
K. esophagus
L. ampulla of Vater

Identify the structures that connect each of the following:

1. _____ Esophagus and duodenum

2. _____ Duodenum and ileum

3. _____ Ascending colon and descending colon

4. _____ Descending colon and rectum

5. _____ Stomach and jejunum

6. _____ Gallbladder and common bile duct

7. _____ Liver and cystic duct

8. _____ Cecum and transverse colon

9. _____ Transverse colon and sigmoid

10. _____ Jejunum and cecum

11. _____ Lower esophageal sphincter (LES) and pyloric sphincter

12. _____ Hepatic flexure and splenic flexure

13. _____ Distal common bile duct and duodenum

14. _____ Pharynx and stomach

READ THE DIAGRAM

Directions: Fill in the blanks with the numbers on the diagram. Not all numbers will be used. See text, pp. 433-449.

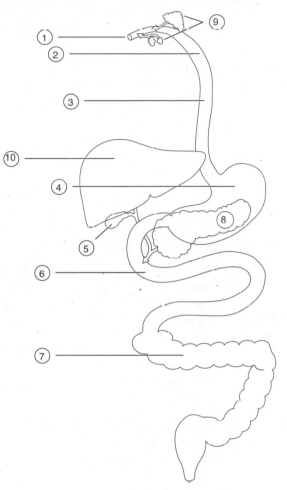

1. _____ Organ that produces bile and secretes it into the hepatic bile ducts

2. _____ The walls of this structure secrete CCK.

3. _____ "Food tube" that carries food from the pharynx to the stomach

4. _____ Pear-shaped sac that attaches to the underside of the liver; concentrates and stores bile

5. _____ Parietal cells of this organ that secrete hydrochloric acid and intrinsic factor

6. _____ Parotid, sublingual, and submandibular

7. _____ Organ that is divided into the fundus, body, and pylorus

8. _____ Most digestion and absorption occur within this structure.

9. _____ The primary function of this organ is to deliver chyme to the duodenum at the proper rate.

10. _____ The walls of this organ are thrown into folds called *rugae*.

11. _____ Divisions include the cecum, colon, rectum, and anal canal.

12. _____ The organ that secretes the most potent digestive enzymes

13. _____ An inflammation of one of these structures is called *mumps*.

14. _____ This organ contains mucous cells, parietal cells, and chief cells.

15. _____ The inner lining of this structure is characterized by villi and microvilli; brush border cells

16. _____ Divisions include the duodenum, jejunum, and ileum.

17. _____ Landmarks of this organ include the greater curvature and lesser curvature.

18. _____ This structure has bends or curves called the *hepatic flexure* and the *splenic flexure*.

19. _____ The pylorus connects the stomach with this structure.

20. _____ Flatus, feces, and defecation are most related to this structure.

21. _____ The chief organ of drug detoxification

22. _____ This organ secretes insulin, glucagon, trypsin, and a bicarbonate-rich secretion.

23. _____ Peristalsis is pendulum-like (sways back and forth) in this organ.

BODY TOON

Hint: A Perfect Ending to the Story of Digestion

Answer: wrecked-em (rectum)

DRAWING AND COLORING

Directions: Using the illustration on the previous page, perform the following exercises.

1. Draw a circle around the LES.

2. Place an *X* over the fundus of the stomach.

3. Draw an arrow to the pylorus.

4. Draw rugae in the stomach wall.

5. Draw an ulcer on the greater curvature of the stomach.

6. Write the word HEPATIC over the proper structure.

7. Draw some stones in the gallbladder.

8. Place a *Y* at the point where the common bile duct empties into the small intestine.

9. Place a *Z* over the area of the appendix.

10. Color the biliary tree *green*.

11. Color the alimentary canal *light blue*.

FROM POINT A TO POINT B

Directions: For each of the following, indicate the structure NOT used as digested food or waste flows from Point A to Point B.

1. Point A: stomach → Point B: cecum

   jejunum   pylorus   duodenum   sigmoid

2. Point A: esophagus → Point B: jejunum

   pyloric sphincter   LES   stomach   cecum

3. Point A: pylorus → Point B: sigmoid

   duodenum   ileum   hepatic flexure   LES

4. Point A: LES → Point B: ileocecal valve

   cecum   stomach   pylorus   duodenum

5. Point A: hepatic flexure → Point B: rectum

   sigmoid   cecum   transverse colon   descending colon

6. Point A: duodenum → Point B: splenic flexure

   cystic duct   cecum   ileocecal valve   jejunum

7. Point A: jejunum → Point B: transverse colon

   duodenum   ileum   ileocecal valve   hepatic flexure

ORDERING

Directions: A piece of food starts in the mouth. Trace the movement of the food from the mouth to the anus (use the words listed below). See text, pp. 435-445.

jejunum	ascending colon
pharynx	hepatic flexure
esophagus	ileocecal valve
transverse colon	descending colon
LES	stomach
sigmoid colon	ileum
duodenum	cecum
rectum	anus

1. mouth

2. _____

3. _____

4. _____

5. _____

6. pyloric sphincter

7. _____

8. _____

9. _____

10. _____

11. _____

12. _____

13. _____

14. _____

15. splenic flexure

16. _____

17. _____

18. _____

19. _____

READ THE DIAGRAM

Liver (Biliary)

Directions: Fill in the blanks with the numbers on the diagram. See text, pp. 445-449.

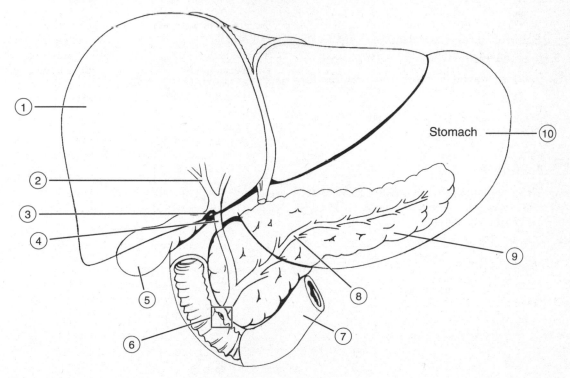

1. _____ The organ that concentrates and stores bile

2. _____ The intestinal segment into which bile empties

3. _____ The common bile duct that empties into the duodenum

4. _____ The duct that delivers bile from the hepatic ducts to the gallbladder

5. _____ This duct merges with the common bile duct at the ampulla of Vater.

6. _____ The organ that produces bile

7. _____ The organ in which bile carries out its digestive function

8. _____ The organ that is inflamed in cholecystitis

9. _____ The structure in which stones accumulate in a person with choledocholithiasis

10. _____ The organ inflamed in hepatitis

11. _____ The organ that empties chyme into the duodenum

12. _____ The organ that contracts and ejects bile in response to CCK

13. _____ The organ that secretes bicarbonate in response to secretin

14. _____ The organ that secretes CCK and secretin in response to chyme

15. _____ The organ removed during a cholecystectomy

16. _____ The organ removed during a gastrectomy

17. _____ The main pancreatic duct that merges with the common bile duct

18. _____ Location of the ampulla of Vater and the sphincter of Oddi

19. _____ Ducts that drain bile from the liver and deliver it to the cystic duct

READ THE DIAGRAM

Directions: Referring to the diagram (Figs. 23.9, 23.10, 23.11) in the textbook, fill in the blanks with the the appropriate figure. See text, pp. 449, 450.

1. _____ The figure that shows the end products of protein digestion

2. _____ The figure that illustrates the hepatic secretion that emulsifies fat globules into tiny fat globules

3. _____ The figure that illustrates the production of glucose, fructose, and galactose

4. _____ The figure that identifies the enzymes that break down carbohydrates into disaccharides

5. _____ The figure that illustrates the result of disaccharide digestion

6. _____ The figure that illustrates the result of trypsin, chymotrypsin, pepsin, and enterokinase digestion

7. _____ The figure that shows the digestion of this substance into amino acids

8. _____ The figure that illustrates the production of sucrose, maltose, and lactose

9. _____ The figure that identifies the enzyme that digests fats to fatty acids and glycerol

10. _____ The figure that identifies an emulsifying agent

11. _____ The figure that describes the digestion of starch to glucose

12. _____ The figure that illustrates the effects of amylases and disaccharidases

MATCHING

Nutrients, Vitamins, and Minerals

Directions: Match the following terms to the most appropriate definition by writing the correct letter in the space provided. See text, pp. 449-455.

A. essential amino acids
B. nonessential amino acids
C. linoleic acid
D. cellulose
E. glucose

F. complete protein
G. incomplete protein
H. saturated fats
I. unsaturated fats
J. vitamins
K. minerals

1. _____ Simplest carbohydrate that is the major fuel used to make ATP

2. _____ Protein that does not contain all the essential amino acids; these include the vegetable proteins, such as nuts, grains, and legumes

3. _____ Amino acids that can be synthesized by the body and do not have to be consumed in the diet

4. _____ A fat that is solid at room temperature; examples include butter and lard

5. _____ Small, organic molecules that help regulate cell metabolism; classified as water-soluble and fat-soluble

6. _____ Complex carbohydrate found primarily in vegetables; provides dietary fiber and bulk to the stools

7. _____ Inorganic substances necessary for normal body function

8. _____ Fat that is liquid at room temperature; usually called *oil*

9. _____ Amino acids that cannot be synthesized by the body and must, therefore, be consumed in the diet

10. _____ Essential fatty acid that is a necessary part of the cell membrane

11. _____ A protein that contains all the essential amino acids; found primarily in animal sources, such as meat, and in some grains and legumes

MATCHING

Vitamins and Minerals

Directions: Match the following terms to the most appropriate definition by writing the correct letter in the space provided. Some terms may be used more than once. See text, pp. 454-456.

A. water-soluble vitamins
B. vitamin A
C. vitamin K
D. iron
E. calcium
F. iodine
G. vitamin C
H. vitamin D
I. fat-soluble vitamins
J. sodium
K. vitamin B group

1. _____ Classification of vitamins A, D, E, K

2. _____ Vitamin necessary for the absorption of calcium and for the development of strong bones

3. _____ A vitamin deficiency that causes night blindness (necessary for the synthesis of rhodopsin)

4. _____ A vitamin deficiency that causes rickets or soft bones (osteomalacia)

5. _____ A vitamin deficiency that causes scurvy, a disease characterized by skin lesions and an inability to heal injured tissue

6. _____ Mineral that helps regulate extracellular volume (including blood volume)

7. _____ Classification of vitamins B and C

8. _____ Mineral necessary for bone growth

9. _____ Mineral necessary for the synthesis of thyroid hormone, which in turn regulates metabolic rate

10. _____ Mineral necessary for the synthesis of hemoglobin and the transport of oxygen

11. _____ Vitamin necessary for the synthesis of prothrombin, a clotting factor

12. _____, _____ Vitamins such as thiamine, niacin, and pyridoxine

13. _____ Deficiency of this vitamin causes bleeding.

SIMILARS AND DISSIMILARS

Directions: Circle the word in each group that is least similar to the others. Indicate the similarity of the three words on the line below each question.

1. duodenum   ileum   cecum   jejunum

_____

2. ileum   fundus   pylorus   body

_____

3. ascending   duodenum   sigmoid   transverse

_____

4. cystic duct   hepatic duct
   common bile duct   pancreatic duct

_____

5. submaxillary   parotid   hepatic   submandibular
                            lobules

_____

6. pyloric   LES   sphincter of Oddi   sigmoid
   sphincter

_____

7. HCl   amylase   trypsin   lipase

_____

8. gastrin   secretin   bile   CCK

_____

9. greater   rugae   villus   lesser curvature
   curvature

_____

191

10. glucose   sucrose   galactose   fructose

_____

11. sucrose   glucose   maltose   lactose

_____

12. trypsin   chymotrypsin   bile   protease

_____

13. lipase   amylase   sucrose   maltase

_____

14. intrinsic factor   HCl   gastrin   ptyalin

_____

15. mucosa   muscular   appendix   serosa

_____

16. trypsin   sucrase   amylase   lactase

_____

17. bile   glucose   lipase   fatty acids and glycerol

_____

18. molars   uvula   incisors   wisdom

_____

19. omentum   peritoneum   mesentery   peristalsis

_____

20. enamel   dentin   pulp   parotid

_____

21. tongue   sublingual glands   parietal cells   frenulum

_____

22. laryngo-   hepatic   naso-   oro-

_____

23. mucous cells   emesis   parietal cells   chief cells

_____

24. satiety center   glucostat theory   eructation   lipostat theory

_____

25. bile   liver   gallbladder   LES

_____

26. ampulla of Vater   common bile duct   ileocecal valve   sphincter of Oddi

_____

27. liver lobule   pylorus   sinusoid   canaliculi

_____

BODY TOON

Hint: The Site of a Musical Performance by the Beatles

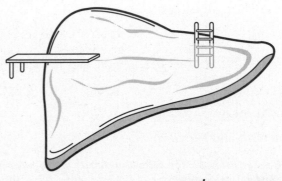

Answer: Liverpool

Part II: Putting It All Together

MULTIPLE CHOICE

Directions: Choose the correct answer.

1. Chyme is
   a. delivered to the duodenum by the stomach.
   b. found in the lacteals.
   c. a pastelike mixture that is stored in the gallbladder.
   d. formed in the mouth in response to ptyalin.

2. The mesentery, omentum, and peritoneal membranes
   a. are located within the alimentary canal.
   b. secrete mucus.
   c. secrete digestive enzymes.
   d. are serous membranes located within the abdominal cavity.

3. Bile is
   a. synthesized by the gallbladder and stored in the liver.
   b. classified as a lipase.
   c. synthesized by the pancreas and stored in the gallbladder.
   d. stored by the gallbladder and released in response to CCK.

4. Why are medications absorbed rapidly when administered sublingually (under the tongue)?
   a. Acid in the saliva dissolves the drugs rapidly.
   b. Ptyalin secreted by the salivary glands digests the drugs, making them more absorbable.
   c. The rich supply of blood vessels under the tongue absorbs the drug.
   d. The warmth of the tongue dissolves the drugs quickly.

5. Parotid, submandibular, and sublingual are words that describe
   a. structures confined to the buccal cavity.
   b. salivary glands.
   c. endocrine glands.
   d. lipase-secreting glands.

6. What nervous statement is true about the GI tract?
   a. The sciatic nerve innervates the descending colon and sigmoid.
   b. Parasympathetic nerve stimulation decreases peristalsis and decreases secretions.
   c. Vagal nerve stimulation decreases peristalsis and decreases secretions.
   d. Blockade of vagal activity slows peristalsis and decreases GI secretions.

7. The act of swallowing directs food from the pharynx into the
   a. larynx.
   b. buccal cavity.
   c. esophagus.
   d. trachea.

8. Gastric reflux or regurgitation causes acidic stomach contents to back up into the
   a. duodenum.
   b. pylorus.
   c. esophagus.
   d. common bile duct.

9. Which of the following is not true of the biliary tree?
   a. includes the hepatic ducts, cystic duct, and common bile duct
   b. carries bile
   c. empties bile into the portal vein
   d. "connects" the liver, gallbladder, and duodenum

10. An amylase
    a. digests fats.
    b. emulsifies fat.
    c. digests carbohydrates.
    d. assembles amino acids into small peptides.

11. If the peristalsis in the large intestine slows,
    a. all waste will be absorbed across the colon wall into the hepatic portal system.
    b. water absorption across the wall of the colon increases, thereby causing a dry or constipated stool.
    c. the person develops diarrhea.
    d. bile refluxes into the main pancreatic duct.

12. What is the primary function of the large intestine?
    a. secretion of potent digestive enzymes
    b. absorption of water and some electrolytes
    c. absorption of the end products of digestion
    d. secretion of CCK

13. CCK
    a. is secreted by the pancreas.
    b. is secreted by the walls of the duodenum in response to the presence of fat.
    c. causes the liver to synthesize lipase.
    d. is a lipase.

14. Which of the following is most likely to result in peritonitis?
    a. gastroesophageal reflux disease (GERD)
    b. cholecystitis
    c. ruptured appendix
    d. pyloric stenosis

15. Peristalsis
    a. occurs only within the stomach and small intestine.
    b. is the same as mastication.
    c. occurs only after deglutition.
    d. moves food along the digestive tract.

16. The LES is
    a. located at the base of the esophagus.
    b. a flap of tissue that covers the glottis and prevents aspiration of food or water.
    c. a piece of tissue that anchors the tongue to the floor of the mouth.
    d. the term that includes the parotid, submaxillary, and submandibular glands.

17. Because hepatic function declines with age, which statement describes an older adult?
    a. likely to be jaundiced
    b. should not eat fat
    c. may require a smaller dosage of a drug
    d. should consume no alcohol

18. The ampulla of Vater and sphincter of Oddi
    a. are at the base of the common bile duct.
    b. control the entrance of chyme into the pylorus.
    c. are part of the LES.
    d. are valvelike structures within the hepatic portal circulation.

193

19. The duodenum and first third of the jejunum are most associated with
    a. deglutition.
    b. secretion of HCl and intrinsic factor.
    c. defecation.
    d. digestion and absorption.

20. Which of the following are found in gastric juice?
    a. bile, secretin, and trypsin
    b. HCl, pepsinogen, and intrinsic factor
    c. HCl, CCK, and ptyalin
    d. trypsinogen, enterokinase, and bile

21. Hematemesis and "coffee-grounds"-appearing vomitus are most associated with
    a. dysphagia and esophageal obstruction.
    b. occlusion of a salivary duct.
    c. bleeding.
    d. common bile duct obstruction.

22. Which of the following organs receives chyme from the stomach and secretions from the liver and pancreas?
    a. cecum
    b. pancreas
    c. pylorus
    d. duodenum

23. Which of the following is least descriptive of the vermiform appendix?
    a. RLQ
    b. cecum
    c. biliary tree structure
    d. lymphatic tissue

24. Which of the following is most descriptive of Peyer's patches?
    a. bile-secreting
    b. trypsin-secreting
    c. lymphatic tissue (MALT)
    d. acid-neutralizing

25. The hepatic sinusoids
    a. are large pore capillaries that contains a mixture of venous blood and arterial blood.
    b. store bile.
    c. are hepatic enzyme-secreting glands.
    d. receive blood from the hepatic veins.

26. Which of the following is most descriptive of trypsin?
    a. hepatic synthesis
    b. disaccharidase
    c. proteolytic enzyme
    d. activated by trypsinogen

27. The presence of dietary fat in the duodenum
    a. suppresses the enteral secretion of CCK.
    b. inhibits the release of bile by the gallbladder.
    c. slows GI motility, thereby enhancing satiety.
    d. suppresses the pancreatic secretion of pancreatic lipase.

28. HCl
    a. is secreted by the gastric parietal cells.
    b. is a proteolytic enzyme.
    c. maintains a gastric pH of 8.
    d. is secreted by Peyer's patches.

CASE STUDY

A.B., a 65-year-old retired nurse, had no previous history of major illness. After a meal of fried chicken and gravy, she complained of moderate to severe midepigastric pain that radiated to the right subscapular region. She became nauseated and vomited. Because there was no improvement in the pain or nausea, she called her physician the following morning. The physician noticed that she was jaundiced and that her stools had become clay colored. She was diagnosed with cholecystitis (inflammation of the gallbladder) and choledocholithiasis (stones in the common bile duct). She was scheduled for a cholecystectomy (surgical removal of the gallbladder) on the following day. One of the preoperative laboratory tests (prothrombin time) indicated that she was hypoprothrombinemic; she was, therefore, given an intramuscular injection of vitamin K.

1. Which of the following best indicates why the fried chicken and gravy stimulated the gallbladder attack?
   a. Fat irritates the gastric lining, causing an excess secretion of hydrochloric acid.
   b. Fat relaxes the LES, causing heartburn.
   c. The presence of fat in the duodenum stimulates the inflamed gallbladder to contract.
   d. The cholesterol in the fats causes stone formation.

2. What causes the jaundice to develop?
   a. There is a hypersecretion of digestive enzymes.
   b. The inflamed gallbladder causes hemolysis.
   c. The inflamed gallbladder secretes a poison that destroys the liver.
   d. The stones block the flow of bile through the common bile duct.

3. Which substance is responsible for the jaundice?
   a. amylase
   b. lipase
   c. hydrochloric acid
   d. bilirubin

4. Why was she hypoprothrombinemic?
   a. The lack of bile in the duodenum decreased the absorption of vitamin K, a factor necessary for the synthesis of prothrombin.
   b. The vomiting caused a vitamin K deficiency; vitamin K is necessary for the synthesis of several clotting factors.
   c. The stones injured the common bile duct, causing it to bleed and depleting the body of clotting factors such as prothrombin.
   d. Prothrombin is synthesized by the gallbladder, so its production decreases because of the inflammation.

5. What is the most likely consequence of hypoprothrombinemia?
   a. jaundice
   b. hypertension
   c. bleeding
   d. hemolysis

6. Vitamin K was given to
   a. stimulate the hepatic synthesis of prothrombin.
   b. dissolve the stones.
   c. stimulate the kidneys to excrete the excess bilirubin.
   d. relieve the pain.

7. Why had her stools become clay colored?
   a. Undigested fat in the duodenum makes the stools appear light.
   b. Because the stones had blocked the common bile duct, there was a decreased amount of bile pigment in the stool.
   c. Excess bile in the duodenum makes the stools appear light.
   d. Small gray stones washed into the stool from the common bile duct, thereby changing its color.

PUZZLE

Hint: Bile Aisle and Main Vein

Directions: Perform the following functions on the Sequence of Words that follows. When all the functions have been performed, you are left with a word or words related to the hint. Record your answer in the space provided.

Functions: Remove the following:

1. The three parts of the small intestine

2. Structure to which the appendix attaches

3. The distal end of the stomach

4. Organ that secretes the most potent digestive enzymes

5. The circular muscle at the distal esophagus

6. Food tube

7. Valve that separates the small intestine from the large intestine

8. Organ that secretes bile

9. Organ that stores bile

10. Four parts of the colon

11. Sphincter at the distal common bile duct

12. Major parasympathetic nerve supplying the gut

13. Disaccharidases (three)

14. A major pancreatic protease

15. Enzyme that digests fats

16. Three salivary glands

17. HCl-secreting gastric cells

18. Landmarks of this organ include the lesser and greater curvatures

Sequence of Words

JEJUNUMSTOMACHGALLBLADDER
SUBMANDIBULARTRANSVERSELO
WERESOPHAGEALSPHINCTERASCE
NDINGCOMMONBILEDUCTPARIETA
LCECUMLIPASEPAROTIDPANCREA
SLACTASEESOPHAGUSSUCRASEPY
LORUSVAGUSSUBLINGUALDESCEN
DINGDUODENUMSIGMOIDPORTAL
VEINLIVERILEUMILEOCECALVALV
EMALTASEODDITRYPSIN

Answer: _____, _____

24 Urinary System

OBJECTIVES

1. List four organs of excretion.
2. Describe the major organs of the urinary system.
3. Describe the location, structure, blood supply, nerve supply, and functions of the kidneys.
4. Explain the role of the nephron unit in the formation of urine.
5. Explain the three processes involved in the formation of urine—filtration, reabsorption, and secretion.
6. Describe the hormonal control of water and electrolytes by the kidneys.
7. List the normal constituents of urine.
8. Describe the structure and function of the ureters, urinary bladder, and urethra.

Part I: Mastering the Basics

MATCHING

The Urinary System

Directions: Match the following terms to the most appropriate definition by writing the correct letter in the space provided. Some terms may be used more than once. See text, pp. 463-465, 474.

A. ureters
B. kidneys
C. hilum
D. renal pelvis
E. urinary bladder
F. renal pyramids
G. renal capsule
H. urethra
I. renal columns
J. calyces

1. _____ Tubes that conduct urine from the kidneys to the bladder

2. _____ Tube that conducts urine from the bladder to the exterior for elimination

3. _____ Reservoir that receives and stores urine

4. _____ A Foley catheter is inserted into this organ for drainage.

5. _____ Bean-shaped organs that make urine

6. _____ Pathogens can ascend from the bladder through these tubes to the kidneys, thereby causing a kidney infection.

7. _____ Structure that is involved in urinary retention

8. _____ Structure that contains the trigone, a triangle formed by the two points of entrance of the ureters and the exit point of the urethra

9. _____ The external sphincter surrounds the upper region of this structure.

10. _____ The internal sphincter is located at the exit of this structure.

11. _____ The prostate gland encircles the proximal end of this structure.

12. _____ The indentation of the bean-shaped kidney; the point where the blood vessels, nerves, and ureter enter or exit the kidney

13. _____ The wall of this structure is arranged in rugae to allow expansion.

14. _____ Basin within the kidney that collects the urine made by the kidney

15. _____ The lighter, outer region of the kidney, called the *renal cortex*, extends inward to form these structures.

16. _____ The darker, inner region of the kidney, called the *renal medulla*, forms these striped cone-shaped regions.

17. _____ Cuplike edges of the renal pelvis that receive the urine from the renal pyramids and empty it into the renal pelvis

18. _____ Tough outer lining that encases the kidney

19. _____ Storage structure that empties during micturition

20. _____ Storage structure that contains the detrusor muscle

Urinary System

Directions: Referring to the illustration, indicate the parts of the urinary system by writing the correct numbers on the lines provided. See text, pp. 463-465, 474.

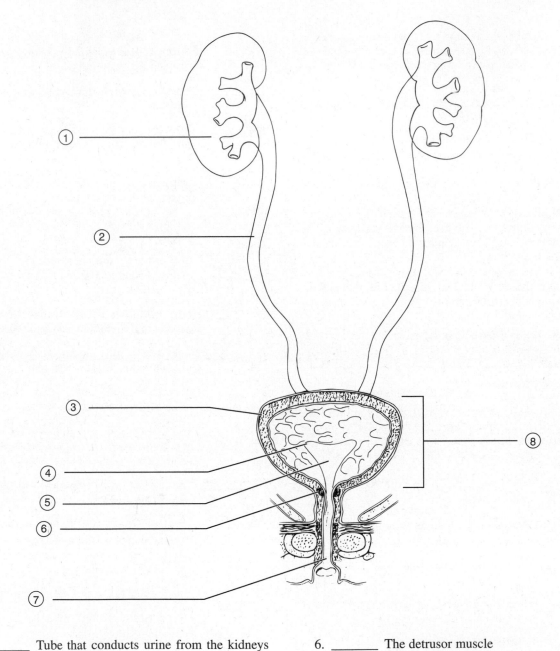

1. _____ Tube that conducts urine from the kidneys to the bladder

2. _____ Urine-making organs

3. _____ Reservoir that stores urine

4. _____ A Foley catheter drains urine from this organ.

5. _____ Pathogens can ascend from the bladder through this tube to the kidney, thereby causing a kidney infection.

6. _____ The detrusor muscle

7. _____ The organ removed during a nephrectomy

8. _____ A triangle formed by the two points of entrance of the ureters and the exit point of the urethra

9. _____ Sphincter at the base of the bladder

10. _____ The ureter enters the bladder at this point.

11. _____ A stone in this structure causes urine to back up into the kidney.

12. _____ Cystitis is an infection of this organ.

13. _____ This structure contracts causing micturition.

14. _____ The organ concerned with urinary retention

15. _____ The organ concerned with renal suppression

16. _____ The proximal end of this structure receives urine; the distal end empties urine into the bladder.

17. _____ This organ receives urine from the ureters and ejects urine into the urethra.

18. _____ This structure is significantly shorter in the female than in the male.

19. _____ The urinary meatus is located on the distal end of this structure.

MATCHING

Nephron Unit

Directions: Match the following terms to the most appropriate definition by writing the correct letter in the space provided. Some terms may be used more than once. See text, pp. 465-468.

A. peritubular capillaries
B. loop of Henle
C. proximal convoluted tubule
D. collecting duct
E. distal convoluted tubule
F. glomerulus (glomeruli)

1. _____ Tuft of capillaries across which water and solute are filtered

2. _____ The distal convoluted tubule empties urine into this structure.

3. _____ Structure that is most concerned with the concentration of urine

4. _____ Vascular structure that surrounds the tubules; involved in reabsorption and secretion

5. _____ The efferent arteriole extends and becomes this structure.

6. _____ Most reabsorption occurs across the walls of this tubular structure.

7. _____ The site at which hormone (ADH) is most active

8. _____ The afferent and efferent arterioles bracket (form bookends to) this structure.

9. _____ The final adjustment of urine occurs at this site.

10. _____ The proximal convoluted tubule extends as this structure.

11. _____ Site at which aldosterone is most active

12. _____ These capillaries sit within the C-shaped Bowman's capsule.

13. _____ Vascular structure that empties blood into the venules and eventually into the renal vein

14. _____ The process of secretion causes solute to move from this structure into the tubules.

15. _____ Composed of an ascending and a descending limb

16. _____ Reabsorption causes water and solute to move from the tubule into this structure.

17. _____ Hairpin structure between the proximal and distal tubules

18. _____ Tubular structure that carries urine from the distal tubule to the calyx

Nephron Unit

Directions: Referring to the diagram, fill in the blanks with the correct number. Some numbers may be used more than once. See text, pp. 465-468.

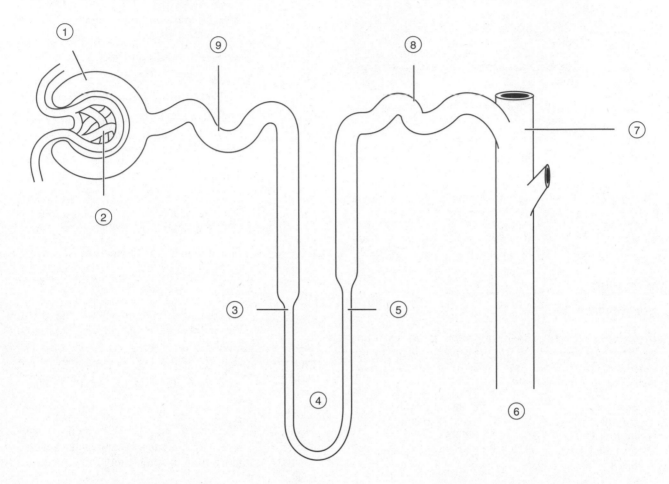

1. _____ Tuft of capillaries across which water and solute are filtered

2. _____ The distal convoluted tubule empties urine into this structure.

3. _____ Structure that is most concerned with the concentration of urine

4. _____ Most reabsorption occurs across the walls of this tubular structure.

5. _____ Site at which ADH is most active

6. _____ Site at which aldosterone is most active

7. _____ The afferent and efferent arterioles "book-end" this structure.

8. _____ The final adjustment of urine occurs at this site.

9. _____ The proximal convoluted tubule extends as this structure.

10. _____ The glomerulus sits within this structure.

11. _____ Composed of an ascending and descending limb

12. _____ Urine on its way to the calyx

13. _____ This structure carries blood.

14. _____ The ascending limb of the loop of Henle

15. _____ The distal convoluted tubule

16. _____ Tubular structure that receives filtrate from Bowman's capsule

COLORING

Directions: Using the illustration above, color the appropriate areas as indicated below.

1. Color the glomerulus *red.*

2. Color the proximal and distal convoluted tubules *green.*

3. Color the loop of Henle *blue.*

4. Color the collecting duct *yellow.*

ORDERING

The Flow of Urine

Directions: Using the words below, trace the flow of urine from its formation (filtration of water and solute across the glomeruli) to its elimination from the body. See text, pp. 465-468, 473-475.

descending limb (loop of   distal convoluted tubule
   Henle)   ureter
collecting duct   calyces
renal pelvis   urinary bladder
urethra   Bowman's capsule
ascending limb (loop of   proximal convoluted
   Henle)   tubule

1. Filtration of water and solute across the glomeruli

2. _____

3. _____

4. _____

5. _____

6. _____

7. _____

8. Calyces

9. _____

10. _____

11. _____

12. _____

CONNECTIONS

Directions: Indicate the structure that forms the connecting link in the following questions. Some letters may be used more than once.

A. proximal convoluted   F. urinary bladder
   tubule   G. peritubular capillaries
B. glomerulus   H. ureter(s)
C. urethra   I. collecting duct
D. descending limb
   (loop of Henle)
E. ascending limb
   (loop of Henle)

1. _____ Connects the proximal convoluted tubule with the ascending limb of the loop of Henle

2. _____ Connects the afferent arteriole with the efferent arteriole

3. _____ Connects the descending limb of the loop of Henle with the distal convoluted tubule

4. _____ Connects the distal convoluted tubule with the calyx

5. _____ Connects the renal pelvis with the urinary bladder

6. _____ Connects Bowman's capsule to the loop of Henle

7. _____ Connects the urinary bladder to the urinary meatus

8. _____ Connects the kidney to the urinary bladder

9. _____ Connects the ureters to the urethra

10. _____ Connects the efferent arteriole to the renal veins

MATCHING

Hormones and Enzymes

Directions: Match the following terms to the most appropriate definition by writing the correct letter in the space provided. Some terms may be used more than once. See text, pp. 468-471.

A. ADH   F. converting enzyme
B. aldosterone   G. PTH
C. renin   H. erythropoietin
D. angiotensin II   I. BNP
E. ANP

1. _____ Hormone that stimulates the adrenal cortex to secrete aldosterone and causes vasoconstriction of the peripheral blood vessels, thereby elevating blood pressure

2. _____ The adrenal cortical hormone that stimulates the distal tubule to reabsorb sodium and excrete potassium

3. _____ Enzyme that changes angiotensin I to angiotensin II

4. _____ The mineralocorticoid that is called the *salt-retaining hormone*

5. _____ Secreted by the posterior pituitary gland; this hormone affects the permeability of the collecting duct to water

6. _____ Released by the kidney in response to hypoxemia; it stimulates red blood cell production by the bone marrow

7. _____ Secreted by the juxtaglomerular apparatus (JGA) when blood pressure or blood volume decreases

8. _____ Secreted by the atrial walls in response to an increase in blood volume; causes the excretion of sodium and water

9. _____ Enzyme that activates angiotensinogen to angiotensin I

10. _____ A deficiency of this hormone causes polyuria (diuresis), sometimes up to 25 L of urine/day; this hormone deficiency disease is called *diabetes insipidus.*

11. _____ Stimulates the renal tubules to reabsorb calcium and to excrete phosphate

12. _____ Hormone that is released in response to ventricular stretch; elevated in heart failure

13. _____ A deficiency of this hormone causes hypocalcemic tetany.

14. _____ Hormone that is deficient in a patient with chronic renal failure; causes anemia

15. _____ ACE inhibitors block the formation of this vasopressor hormone.

16. _____ A deficiency of this hormone causes the urinary excretion of Na^+ and water and the retention of K^+.

17. _____ An excess of this hormone can cause hypercalcemia and kidney stones.

TELL A STORY

It's a Blood Pressure Thing!

Directions: Using the words below, fill in the blanks to complete the story.

blood pressure	K^+
aldosterone	vasopressor
liver	Na^+
angiotensin II	angiotensinogen
JGA cells	converting enzyme
distal tubule	water

The story starts with an inactive hormone called angiotensinogen that is secreted into the blood by the _____. In response to a low blood pressure or concentrated blood, as in dehydration, renin is secreted by _____ of the kidneys. Renin activates _____ to angiotensin I. _____ then activates angiotensin I to _____. This hormone travels by way of the blood to the adrenal cortex, where it stimulates the release of _____. This salt-retaining hormone travels by way of the blood to the _____ of the nephron, causing it to reabsorb _____ and _____ and to excrete

_____. Angiotensin II is also a vasoconstrictor, called a _____ agent, and therefore increases blood pressure. Thus, activation of the renin-angiotensin-aldosterone system expands blood volume and increases systemic vascular resistance, thereby increasing _____.

TELL A STORY

Going Up!

Directions: Using the words below, fill in the blanks to complete the story.

pyelonephritis	pyuria
analgesia	cystitis
antimicrobial	ascending infection
dysuria	hematuria

Polly Uria is a college freshman who came to the student health office complaining of flank pain, night sweats, fever, and "feeling terrible." She reported that she had been experiencing frequency, urgency, and pain on urination (called _____) for 1 week before the onset of fever and pain. She did not seek medical treatment at this time because she was busy with exams. Instead, she had self-medicated with an over-the-counter azo dye preparation. The azo dye relieved the pain; this effect is called _____. The azo dye, however, does not kill the pathogens and therefore does not exert a(n) _____ effect. A cloudy and foul-smelling urine sample indicated pus and bacteria in the urine (called _____) and blood in the urine (called _____). She was diagnosed as having a kidney infection (called _____) that had developed in response to an untreated bladder infection (called _____). Treatment included an antibiotic, forced fluids, and bed rest. The development of a kidney infection from an untreated bladder infection is an example of a(n) _____ because the infection moved up from the bladder to the kidney. Moral of the story: Do NOT ignore a bladder infection!

SIMILARS AND DISSIMILARS

Directions: Circle the word in each group that is least similar to the others. Indicate the similarity of the three words on the line below each question.

1. cortex   medulla   trigone   renal columns

   _____

2. afferent   ureter   peritubular   glomeruli
   arteriole   capillaries

   _____

3. proximal tubule   loop of Henle   urinary bladder   collecting duct

_____

4. trigone   renal pelvis   detrusor muscle   urinary bladder

_____

5. aldosterone   ADH   glucosuria   BNP

_____

6. renal vein   renal artery   peritubular capillaries   loop of Henle

_____

7. angiotensin II   aldosterone   converting enzyme   azotemia

_____

8. ascending limb   calyx   descending limb   loop of Henle

_____

9. pyuria   glucosuria   dialysis   hematuria

_____

10. polyuria   oliguria   diuresis   detrusor

_____

11. urination   azotemia   micturition   voiding

_____

12. uremia   renal failure   cystitis   azotemia

_____

13. 125 mL/min   180 L/24 hr   renal secretion   GFR

_____

14. collecting duct   glomeruli   loop of Henle   renal tubules

_____

15. nephron unit   detrusor muscle   Bowman's capsule   peritubular capillaries

_____

16. azotemia   glucosuria   polyuria   pyuria

_____

Part II: Putting It All Together

MULTIPLE CHOICE

Directions: Choose the correct answer.

1. ADH
   a. Is secreted by the adrenal cortex.
   b. increases the reabsorption of water by the collecting duct membrane.
   c. is the salt-retaining hormone.
   d. increases the glomerular filtration rate (GFR).

2. Why is glucose not normally found in the urine?
   a. No glucose is filtered by the glomerulus.
   b. The filtered glucose is used by the kidney cells for energy.
   c. All filtered glucose is reabsorbed.
   d. The filtered glucose is converted to creatinine and then excreted.

3. Which condition is most likely to cause glucose to be excreted in the urine?
   a. hypertension
   b. renal failure
   c. hyperglycemia
   d. jaundice

4. Which of the following is least true of the urinary bladder?
   a. It contains the detrusor muscle.
   b. It receives urine from two ureters.
   c. The bladder wall is responsive to the effects of ADH.
   d. Micturition occurs when the detrusor muscle contracts and the sphincters relax.

5. When GFR decreases,
   a. diuresis occurs.
   b. aldosterone secretion decreases.
   c. ADH secretion diminishes.
   d. urinary output declines.

6. Which of the following is a consequence of a diminished GFR?
   a. glucosuria
   b. polyuria
   c. oliguria
   d. hematuria

7. What happens at the glomerular membrane?
   a. Water and dissolved solute are filtered into Bowman's capsule.
   b. The JGA cells release ADH.
   c. The JGA cells release aldosterone.
   d. The glomerular membrane reabsorbs Na^+.

8. Which of the following is true of the distal tubule?
   a. receives urine from the collecting duct
   b. responds to aldosterone
   c. nephron structure that is primarily concerned with filtration
   d. also called *Bowman's capsule*

9. Which of the following is not a function of aldosterone?
   a. causes Na^+ reabsorption at the distal tubule
   b. exerts a kaliuretic effect
   c. causes the reabsorption of K^+ by the collecting duct
   d. expands blood volume

10. Which of the following is the most accurate statement regarding creatinine?
    a. All filtered creatinine is reabsorbed.
    b. No creatinine is filtered.
    c. Most filtered creatinine is eliminated in the urine.
    d. Creatinine may be secreted but is never filtered.

11. An elevated serum creatinine level is most indicative of
    a. cystitis.
    b. renal calculi.
    c. kidney failure.
    d. urinary retention.

12. Which of the following is a true statement?
    a. Most water reabsorption occurs at the proximal tubule.
    b. Most filtration occurs across the walls of the collecting duct.
    c. The glomerular membrane is sensitive to both ADH and aldosterone.
    d. Collecting duct function is regulated by ANP and BNP.

13. When the renin-angiotensin-aldosterone system is activated,
    a. the JGA cells secrete angiotensinogen.
    b. angiotensin I causes the adrenal cortical secretion of ADH.
    c. angiotensin II causes the adrenal cortical secretion of aldosterone.
    d. aldosterone increases the renal excretion of sodium and water.

14. Bowman's capsule receives water and dissolved solute from the
    a. proximal convoluted tubule.
    b. glomerulus.
    c. collecting duct.
    d. renal pelvis.

15. The efferent arteriole
    a. receives urine from the collecting duct.
    b. delivers blood to the peritubular capillary.
    c. delivers blood to the glomerulus for filtration.
    d. secretes converting enzyme.

16. The renal pelvis
    a. is part of the nephron unit.
    b. receives urine from the calyces.
    c. receives blood from the peritubular capillaries.
    d. concentrates the urine as it leaves the nephron unit.

17. Angiotensin II
    a. is secreted by the JGA cells.
    b. is produced in response to the activity of converting enzyme.
    c. is a powerful vasodilator.
    d. prevents the release of aldosterone.

18. Drugs (e.g., atropine) that interfere with the relaxation of the urinary bladder sphincter are most likely to cause
    a. renal failure.
    b. renal calculi.
    c. urinary retention.
    d. diuresis.

19. A person with a stenosed (narrowed) renal artery is most likely to present with
    a. urinary retention.
    b. uremia.
    c. hypertension.
    d. hematuria.

20. A person with damaged glomeruli filters large amounts of albumin and therefore develops
    a. severe hypertension.
    b. a kidney infection.
    c. generalized edema.
    d. cystitis.

21. The development of uremia suggests
    a. a bladder infection.
    b. urinary retention.
    c. glucosuria and osmotic diuresis.
    d. renal failure.

22. Which of the following renal responses occurs when the arterial blood pressure declines to 70/50 mm Hg?
    a. The collecting duct becomes unresponsive to ADH.
    b. GFR declines.
    c. Diuresis occurs.
    d. Na$^+$ excretion increases.

23. Which of the following is correct?
    a. two urinary bladders
    b. two urethras
    c. two ureters
    d. one kidney

24. The nephron units
    a. are the urine-making structures of the kidney.
    b. line the urinary tract.
    c. are found within the urinary bladder.
    d. are responsible for micturition.

25. Which of the following is formed by the entrance and exit of the ureters and urethra into/out of the urinary bladder?
    a. JGA
    b. calyces
    c. trigone
    d. detrusor

26. Kaliuresis
    a. is a response to a deficiency of aldosterone.
    b. is a response to a deficiency of ADH.
    c. is a consequence of renal failure.
    d. refers to the urinary excretion of K$^+$.

27. The loop of Henle
    a. is the nephron structure across which most filtration occurs.
    b. is part of the renal tubular structures of the nephron unit.
    c. receives urine from the collecting duct.
    d. is part of the vascular structures of the nephron units.

28. The internal and external sphincters are most associated with
    a. micturition.
    b. kaliuresis.
    c. GFR.
    d. reabsorption of glucose.

29. The reabsorption of Na$^+$
    a. generally determines the reabsorption of water.
    b. is determined primarily by ADH.
    c. occurs primarily within the trigone.
    d. is determined by PTH and accompanied by the reabsorption of Ca^{2+}.

CASE STUDY

Di Uresis is a college freshman who came to the student health office complaining of pain on urination accompanied by frequency and urgency. A urine sample indicated the presence of pathogens, pus, and blood. She had no fever and no evidence of flank pain or night sweats. She was diagnosed as having cystitis and was given an antibiotic and instructed to force fluids and rest.

1. Which part of the lower urinary tract was inflamed?
    a. prostate gland
    b. renal pelvis
    c. urinary bladder
    d. nephron unit

2. Which of the following explains why the physician was concerned that Di might develop a kidney infection?
    a. The renal artery carries infected blood to the kidney.
    b. The pathogens in the bladder can ascend (crawl up) the ureters to infect the kidneys.
    c. She might stop eating, thereby becoming more susceptible to infection.
    d. The high bacterial count in her urine would suppress her immune system.

3. Which of the following statements explains why cystitis is more common in women than in men?
    a. The urethra is shorter in the female than in the male.
    b. Women excrete a less acidic urine than men do.
    c. Women excrete more glucose in the urine than men do.
    d. Men urinate more frequently than women do.

4. Which statement about urine is true?
    a. Urine is normally colonized by *Escherichia coli*.
    b. Urinary tract infection is more common in acidic urine than in alkaline urine.
    c. Urine is normally sterile.
    d. Urine normally contains glucose, protein, and pus.

Hint: "Scanty and Not So Scanty"

Directions: Perform the following functions on the Sequence of Words that follows. When all the functions have been performed, you are left with a word or words related to the hint. Record your answer in the space provided.

Functions: Remove the following:

1. The structure across which 180 L of water is filtered every day

2. The salt-retaining hormone that is secreted by the adrenal cortex

3. The structure that contains the ascending and descending limbs

4. The name of the capillaries most concerned with reabsorption

5. The structure that is most affected by ADH

6. The cation that is excreted in response to aldosterone

7. The tube that carries urine from the renal pelvis to the bladder

8. The smooth muscle in the urinary bladder; contraction causes micturition

9. The process of urinating or voiding

10. Inhibition of the reabsorption of this cation causes diuresis.

11. Three forms of nitrogenous waste

12. Secreted by the JGA in response to a decline in blood pressure

13. Organ that temporarily stores urine

14. Gland that secretes ADH and the gland that secretes aldosterone

15. That which activates angiotensin I to angiotensin II

Sequence of Words

A L D O S T E R O N E A M M O N I A M I C T U R I T
I O N C O N V E R T I N G E N Z Y M E P O T A S S I U
M G L O M E R U L U S L O O P O F H E N L E C R E A
T I N I N E D E T R U S O R U R I N A R Y B L A D D E
R O L I G U R I A N E U R O H Y P O P H Y S I S C O L
L E C T I N G D U C T R E N I N S O D I U M P O L Y U
R I A U R E A P E R I T U B U L A R U R E T E R A D R
E N A L C O R T E X D I U R E S I S

Answer: _____, _____,
_____

25 Water, Electrolyte, and Acid–Base Balance

Answer Key: Textbook page references are provided as a guide for answering these questions. A complete answer key was provided for your instructor.

OBJECTIVES

1. Describe the two main fluid compartments and the composition of body fluids.
2. Define *intake* and *output*.
3. Explain the effects of water imbalances, fluid shift, and fluid spacing.
4. List factors that affect electrolyte balance.
5. Describe the most common ions found in the intracellular and extracellular compartments.
6. List three mechanisms that regulate pH in the body.
7. Discuss acid–base imbalances: acidosis and alkalosis.

Part I: Mastering the Basics

MATCHING

Fluid Compartments

Directions: Match the following terms to the most appropriate definition by writing the correct letter in the space provided. Some terms may be used more than once. See text, pp. 482, 483.

A. intracellular   D. transcellular
B. interstitial   E. lymph
C. plasma

1. _____ Water within blood vessels

2. _____ Includes cerebrospinal fluid, the aqueous and vitreous humors in the eye, synovial fluids of joints, serous fluids within body cavities, and glandular secretions

3. _____ Most water (about 63%) is located in this compartment.

4. _____ Water located between the cells

5. _____ *Hypervolemia* refers to an increase in this water compartment.

6. _____ Also called *tissue fluid*

7. _____ Largest extracellular fluid compartment

8. _____ Fluid of this compartment empties into the subclavian veins.

9. _____ Poor skin turgor is caused by a decrease of fluid within this extracellular compartment.

Directions: Referring to the diagram, fill in the blanks with the correct number that corresponds to the fluid compartments and their divisions. See text, pp. 482, 483.

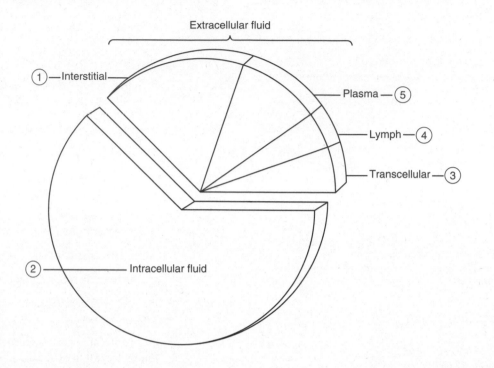

1. _____ Water within blood vessels

2. _____ Includes cerebrospinal fluid, the aqueous and vitreous humors in the eye, synovial fluids of joints, serous fluids within body cavities, and glandular secretions

3. _____ Most water (about 63%) is located in this compartment.

4. _____ Water located between the cells

5. _____ *Hypervolemia* refers to an increase in this water compartment.

6. _____ Also called *tissue fluid*

7. _____ Largest extracellular fluid compartment

8. _____ Fluid of this compartment empties into the subclavian veins.

9. _____ Poor skin turgor is caused by a decrease of fluid within this extracellular compartment

MATCHING

Major Ions

Directions: Match the following terms to the most appropriate definition by writing the correct letter in the space provided. Some terms may be used more than once. See text, pp. 485-487.

A. sodium (Na^+)   D. bicarbonate (HCO_3^-)
B. potassium (K^+)   E. calcium (Ca^{2+})
C. hydrogen (H^+)   F. chloride (Cl^-)

1. _____ Plasma concentration of this ion determines pH.

2. _____ Chloride usually follows the movement of this cation.

3. _____ Chief extracellular cation

4. _____ Most diuretics work by blocking the renal reabsorption of this cation.

5. _____ More than 99% of this cation is stored in the bones and teeth.

6. _____ Neutralizes a base

7. _____ Chief intracellular cation

8. _____ *Hypernatremia* and *hyponatremia* refer to abnormal plasma concentrations of this ion.

9. _____ *Hyperkalemia* and *hypokalemia* refer to abnormal plasma concentrations of this ion.

10. _____ Many diuretics cause kaliuresis, thereby causing a loss of this ion.

11. _____ *Hypercalcemia* and *hypocalcemia* refer to abnormal plasma concentrations of this ion.

12. _____ Aldosterone causes the kidneys to excrete this ion.

13. _____ The chief extracellular anion

14. _____ An increase in this cation causes acidosis.

15. _____ Regulated primarily by parathyroid hormone (PTH)

16. _____ Aldosterone stimulates the kidney to reabsorb this cation.

17. _____ The respiratory system is most sensitive to the plasma concentrations of this cation.

18. _____ Carbon dioxide (CO_2) is transported in the blood in this form.

19. _____ Buffers H^+

MATCHING

Acids and Bases

Directions: Match the following terms to the most appropriate definition by writing the correct letter in the space provided. The terms may be used more than once. See text, pp. 487-491.

A. acidosis   B. alkalosis

1. _____ Acid–base imbalance associated with a plasma pH lower than 7.35

2. _____ Acid–base imbalance associated with a plasma pH higher than 7.45

3. _____ Acid–base imbalance that develops when the plasma concentration of H^+ decreases

4. _____ This acid–base imbalance develops when the plasma concentration of H^+ increases.

5. _____ Describes an imbalance characterized by an increase in H^+ concentration and a decrease in pH

6. _____ Describes an imbalance characterized by a decrease in H^+ concentration and an increase in pH

7. _____ Rapid, incomplete catabolism of fatty acids produces ketone bodies and causes this acid–base imbalance.

8. _____ Kussmaul respirations help correct this acid–base imbalance.

9. _____ Emphysema, high doses of narcotics, and splinting the chest may all cause this acid–base imbalance.

10. _____ An anxious hyperventilating patient is most likely to develop this acid–base imbalance.

11. _____ Any condition that causes hyperventilation causes this acid–base imbalance.

12. _____ Any condition that causes hypoventilation causes this acid–base imbalance.

13. _____ Vomiting of stomach contents is most likely to cause this acid–base imbalance.

TELL A STORY

All Wet

Directions: Using the words below, fill in the blanks to complete the story. See text, pp. 483-485.

cyanosis	dyspnea
K^+	water
orthopnea	Na^+
diuresis	

Anna Sarca awoke suddenly at 3 AM experiencing difficulty breathing (called _____). In fact, she could only breathe when she was sitting in an upright position (called _____). Her lips were blue (called _____), and she was coughing up a pink frothy fluid. She was rushed to the ER and was diagnosed with acute heart failure and pulmonary edema. She was given digoxin to strengthen her heart and furosemide (Lasix), a water pill. Oxygen by mask helped relieve the respiratory distress. The furosemide acts on the renal tubules to block the reabsorption of sodium (symbol, _____). As a result, _____ reabsorption is also decreased, resulting in an increase in urinary output (called _____). Unfortunately, furosemide also causes kaliuresis, the loss of _____ in the urine. Anna was soon doing better and was discharged in 1 week. Discharge medications included digoxin (Lanoxin), a thiazide diuretic, and a potassium supplement.

SIMILARS AND DISSIMILARS

Directions: Circle the word in each group that is least similar to the others. Indicate the similarity of the three words on the line below each question.

1. plasma   transcellular   intracellular   interstitial

_____

2. ion   precipitate   anion   cation

_____

209

3. hydrogen ion   sodium ion   chloride   potassium
                                ion   ion

_____

4. K^+ hyperkalemia   hyponatremia   hypokalemia

_____

5. tenting   dehydration   Kussmaul   poor skin
                                  respirations   turgor

_____

6. pH   hyperuricemia   acidosis   alkalosis

_____

7. Ca^{2+} osteoclastic activity   PTH   respiratory
                                         acidosis

_____

8. pH 7.8   increased $[H^+]$ acidosis   pH 7.2

_____

9. alkalosis   decreased $[H^+]$ pH 7.6   pH 7.2

_____

10. hyperkalemia   kaliuresis

     hypocalcemia   potassium

_____

11. hypoventilation   acidosis   pH 7.3   hypocalcemia

_____

Part II: Putting It All Together

MULTIPLE CHOICE

Directions: Choose the correct answer.

1. Which of the following is true of water balance?
   a. Thirst is the primary regulator of water intake.
   b. Most water is eliminated by the kidneys.
   c. Water can be lost through sensible and insensible water loss.
   d. All of the above are true.

2. What is the average adult intake of water in 24 hours?
   a. 2500 mL
   b. 500 mL
   c. 1000 mL
   d. 180 L

3. Which of the following organs are the primary regulators of water output?
   a. lungs
   b. skin and accessory organs
   c. kidneys
   d. organs of the digestive tract

4. ADH
   a. is released by the posterior pituitary gland when the blood volume decreases.
   b. is the salt-retaining hormone.
   c. is a mineralocorticoid.
   d. stimulates the glomerulus to filter water.

5. Aldosterone
   a. determines the permeability of the collecting duct to water.
   b. is a neurohypophyseal hormone.
   c. stimulates the reabsorption of K^+ by the distal tubule.
   d. is the salt-retaining hormone.

6. Which type of drug is most likely to cause diuresis?
   a. a drug that mimics the effects of ADH
   b. a drug that mimics the effects of aldosterone
   c. a drug that reabsorbs bicarbonate
   d. a drug that blocks the tubular reabsorption of Na^+

7. Water usually follows the movement of which ion?
   a. K^+
   b. Ca^{2+}
   c. Na^+
   d. HCO_3^-

8. Poor skin turgor is
   a. indicative of blood volume expansion and heart failure.
   b. treated with antibiotics.
   c. indicative of dehydration.
   d. the earliest symptom of respiratory acidosis.

9. What does pH measure?
   a. hydration status
   b. H^+ concentration $[H^+]$
   c. hemolysis
   d. urine concentration

10. A buffer
    a. adds only H^+.
    b. adds only bicarbonate ion.
    c. prevents changes in pH.
    d. stimulates Kussmaul respirations.

11. What is the normal range for blood pH?
    a. 7.00 to 7.35
    b. 7.35 to 7.55
    c. 7.35 to 7.45
    d. 6.85 to 7.45

12. Which of the following is the first line of defense in the regulation of acid–base balance?
    a. respiratory system
    b. kidneys
    c. integumentary system
    d. buffers

13. Which of the following is not a function of aldosterone?
    a. causes Na^+ reabsorption at the distal tubule
    b. exerts a kaliuretic effect
    c. causes the reabsorption of K^+ by the collecting duct
    d. expands blood volume

14. Which of the following maintains pH by donating H^+ or removing H^+?
    a. PTH/ADH
    b. ADH/ANP
    c. HCO_3^-/H_2CO_3
    d. Na^+/K^+

15. Why is an older adult more prone to dehydration?
    a. The collecting duct becomes overresponsive to the effects of ADH.
    b. The adrenal cortex oversecretes aldosterone.
    c. There is a decrease in the thirst mechanism.
    d. The person retains excessive sodium.

16. Which of the following systems helps correct metabolic acidosis by blowing off CO_2?
    a. renal system
    b. respiratory system
    c. baroreceptors
    d. lymphatic system

17. Why do patients in kidney failure develop acidosis? The kidneys
    a. eliminate too much water.
    b. cannot eliminate H^+.
    c. cannot eliminate K^+.
    d. make excess bicarbonate.

18. Why do patients in kidney failure become hyperkalemic? The kidneys
    a. eliminate too much water.
    b. cannot eliminate calcium.
    c. cannot eliminate K^+.
    d. fail to secrete erythropoietin.

19. In addition to buffering, how does the body of a patient in diabetic ketoacidosis try to correct the acid–base imbalance?
    a. The kidneys reabsorb all filtered H^+.
    b. The kidneys excrete bicarbonate (HCO_3^-).
    c. The rate and depth of respirations increase (Kussmaul respirations).
    d. The kidneys shut down to conserve water.

20. What does urinary specific gravity measure?
    a. the ability of the kidney to filter protein
    b. glomerular filtration rate (GFR)
    c. how much glucose is in the urine
    d. the concentration of urine

21. A kaliuretic effect is achieved when
    a. Na^+ is reabsorbed.
    b. Ca^{2+} is excreted in the urine.
    c. blood volume expands.
    d. potassium is excreted in the urine.

22. A patient is admitted with severe emphysema and a PO_2 of 80 mm Hg. He has a blood pH of 7.25 and a serum bicarbonate level of 40 mEq/L. Which of the following is an accurate description?
    a. metabolic acidosis and respiratory compensation
    b. metabolic alkalosis with respiratory compensation
    c. respiratory alkalosis with renal compensation
    d. respiratory acidosis with renal compensation

23. A diabetic patient was admitted to the ER with an infection and acid–base imbalance. The patient has admitted to not taking his insulin for the past several days because "I wasn't feeling well and not eating much." Laboratory tests indicate a blood pH of 7.25, a low plasma bicarbonate level, low PCO_2, and elevated PO_2. Which statement is not true?
    a. He has metabolic acidosis with respiratory compensation.
    b. He is most likely hyperglycemic.
    c. He has respiratory acidosis with compensatory Kussmaul respirations.
    d. His laboratory values are consistent with ketoacidosis.

24. Which group is incorrect?
    a. cations: sodium, potassium, calcium
    b. acid–base imbalances: acidosis, alkalosis
    c. lines of defense against acid–base imbalance: buffers, lungs, kidneys
    d. transcellular fluids: aqueous humor, cerebrospinal fluid, synovial fluid, plasma

25. Which group is incorrect?
    a. transcellular fluids: aqueous humor, cerebrospinal fluid, synovial fluid
    b. buffers: bicarbonate/carbonic acid, phosphate, hemoglobin, plasma proteins
    c. lines of defense against acid–base imbalance: buffers, lungs, kidneys
    d. anions: bicarbonate, chloride, sodium

26. Which group is incorrect?
    a. cations: sodium, potassium, calcium
    b. anions: bicarbonate, chloride
    c. acid–base imbalances: acidosis, alkalosis
    d. words that relate to potassium: kaliuresis, hypokalemia, hypernatremia

CASE STUDY

A.S. is a 65-year-old retired engineer. He had a myocardial infarction (heart attack) 5 years ago and has done well since his recovery. Two days ago, he noticed that he became short of breath while walking. He is unable to sleep while lying flat and now requires three pillows. His physician determined that A.S. was in heart failure and that his lungs were "wet." He was prescribed a strong diuretic, furosemide (Lasix), and digoxin (to strengthen his cardiac muscle contraction). He was also given a potassium supplement.

1. A.S.'s "wet" lungs and the inability to sleep while lying flat are indications of which condition?
   a. acidosis
   b. dehydration
   c. pulmonary edema
   d. emphysema

2. Why was the diuretic administered?
   a. to remove the excess fluid
   b. to strengthen the cardiac muscle contraction
   c. to heal the heart muscle
   d. to buffer H^+

3. How do most diuretics work?
   a. by increasing the elimination of H^+
   b. by blocking the tubular reabsorption of Na^+
   c. by making the collecting duct unresponsive to ADH
   d. by decreasing the glomerular filtration of water

4. How does the heart drug digoxin (+ inotropic agent) increase urinary output?
   a. It increases blood flow to the kidney, thereby increasing the amount of water filtered by the kidney.
   b. It blocks Na+ reabsorption by the kidney.
   c. It enhances the urinary excretion of K^+.
   d. It antagonizes the effects of furosemide.

5. Why was the potassium administered?
   a. to heal the injured myocardium
   b. to replace K^+ lost because of the kaliuretic effect of the diuretic
   c. to enhance the tubular reabsorption of Na^+
   d. to buffer H^+

6. What kind of dietary modification will be prescribed?
   a. increased calcium
   b. forced fluids
   c. sodium restriction
   d. high calorie intake

PUZZLE

Hint: Who's All Wet?

Directions: Perform the following functions on the Sequence of Words that follows. When all the functions have been performed, you are left with a word or words related to the hint. Record your answer in the space provided.

Functions: Remove the following:

1. Four extracellular fluid spaces

2. Ions affected in hyperkalemia, hypernatremia, and hypercalcemia

3. Organ that eliminates most water from the body

4. Anion that follows the active pumping of sodium

5. pH imbalance caused by hypoventilation

6. Any element that carries an electrical charge

7. Clinical state caused by a deficiency of water

8. pH imbalance caused by vomiting of stomach contents

9. $NaCl \rightarrow Na^+ + Cl^-$

10. Tries to maintain normal blood pH; example is $HCO_3^- / NaHCO_3$

11. Neurohypophyseal hormone that affects the permeability of the collecting duct to water

12. Adrenal cortical hormone that is the "salt-retaining" hormone

13. The steroidal classification of the "salt-retaining" hormone

14. Respiratory compensation for metabolic acidosis

Sequence of Words

```
TRANSCELLULARMINERALOCORT
ICOIDSODIUMKIDNEYACIDOSISA
NASARCAADHCHLORIDELYMPHPO
TASSIUMALDOSTERONEIONINTER
STITIALKUSSMAULRESPIRATION
SDEHYDRATIONIONIZATIONALKA
LOSISBUFFERGENERALIZEDEDEM
ACALCIUMPLASMA
```

Answer: _____

Hint: A Positively Charged Ion

Answer: cation

26 Reproductive Systems

Answer Key: Textbook page references are provided as a guide for answering these questions. A complete answer key was provided for your instructor.

OBJECTIVES

1. List and describe the structures and functions of the male reproductive system.
2. Describe the hormonal control of male reproduction, including the effects of testosterone.
3. List and describe the structures and functions of the female reproductive system.
4. Explain the hormonal control of the female reproductive cycle and the two reproductive cycles.
5. Describe the structure of the breast and lactation
6. Describe the various methods of birth control.

Part I: Mastering the Basics

MATCHING

Reproductive Cells and Structures

Directions: Match the following terms to the most appropriate definition by writing the correct letter in the space provided. Some terms may be used more than once. See text, pp. 496-500.

A. ovaries
B. testes
C. ovum
D. sperm
E. spermatogonia
F. seminiferous tubules
G. interstitial cells
H. epididymis
I. penis
J. ejaculatory duct
K. prepuce
L. external genitals
M. scrotum
N. urethra

1. _____ Male gonads; also called *testicles*

2. _____ Cells that secrete testosterone

3. _____ Mature male gamete

4. _____ Female gonads

5. _____ The vas deferens curves and joins with the duct of the seminal vesicle to form this duct.

6. _____ Sac or pouch located between the thighs; holds the testicles

7. _____ Structure that produces sperm and secretes testosterone

8. _____ Name for the penis and the scrotum

9. _____ Undifferentiated spermatogenic cells

10. _____ Coiled structure that sits on top of the testes; a place where sperm mature

11. _____ Tubular structure that carries both urine and semen

12. _____ Tubular structure that extends from the bladder to the tip of the penis

13. _____ Male copulatory organ

14. _____ Composed of a head, body (midpiece), and tail

15. _____ Two ejaculatory ducts pass through the prostate gland and join with this urinary structure.

16. _____ Female gamete

17. _____ Piece of skin that is circumcised; also called the *foreskin*

18. _____ Skin, skin, skin, skin

MATCHING

Hormones, Glands, and Semen

Directions: Match the following terms to the most appropriate definition by writing the correct letter in the space provided. Some terms may be used more than once. See text, pp. 500-504.

A. semen
B. prostate
C. emission
D. seminal vesicles
E. orgasm
F. impotence
G. testosterone
H. erection
I. luteinizing hormone (LH)
J. releasing hormones

1. _____ The inability to achieve an erection; also called *erectile dysfunction*

2. _____ Caused by the filling of the erectile tissue by blood

3. _____ Pleasurable sensation experienced during ejaculation

4. _____ The penis enlarges and becomes rigid.

5. _____ An androgen necessary for sperm development and for the male secondary sex characteristics

6. _____ Movement of sperm and glandular secretions from the testes and the genital ducts into the urethra

215

7. _____ The hypothalamus secretes these hormones, which in turn stimulate the anterior pituitary gland to release gonadotropins.

8. _____ Gland that encircles the upper part of the urethra; when enlarged, it impairs urination (dysuria)

9. _____ Gland that secretes 60% of the semen

10. _____ Also called *interstitial cell–stimulating hormone* (ICSH); stimulates the interstitial cells to secrete testosterone

11. _____ Mixture of sperm and the secretions of the accessory glands

12. _____ Walnut-shaped gland that surrounds the proximal urethra as it leaves the bladder; this gland contributes to the formation of semen

ORDERING

Directions: Place these structures in order by writing the correct term in the space provided. The first one is given. See text, pp. 497, 498.

vas deferens   epididymis
ejaculatory duct   urethra

1. seminiferous tubules

2. _____

3. _____

4. _____

5. _____

MATCHING

Structures of the Female Reproductive Tract

Directions: Match the following terms to the most appropriate definition by writing the correct letter in the space provided. Some terms may be used more than once. See text, pp. 502-506.

A. ovaries   G. follicular cells
B. oocyte   H. corpus luteum
C. vagina   I. graafian follicle
D. uterus   J. corpus albicans
E. perineum   K. hymen
F. fallopian tubes L. external genitals

1. _____ The corpus luteum dies and becomes this nonsecreting structure, also known as the *white body.*

2. _____ Skin-covered muscular region between the vaginal orifice and the anus

3. _____ Implantation of the embryo occurs here, also called the *womb.*

4. _____ The growing embryo and fetus live here for 9 months.

5. _____ The placenta develops in this organ.

6. _____ Contains the infundibulum and fimbriae; transports the ovum to the uterus

7. _____ Fertilization occurs here.

8. _____ Pear-shaped organ that is held in place by the broad ligaments

9. _____ Mons pubis, labia, and clitoris; also called the *vulva*

10. _____ Parts of this organ are the fundus, body, and cervix.

11. _____ Layers of this organ are the epimetrium (perimetrium), myometrium, and endometrium.

12. _____ 4-inch muscular tube that extends from the cervix to the vaginal opening on the perineum

13. _____ Thin membrane that partially covers the vaginal opening

14. _____ Female gonads

15. _____ Immature egg

16. _____ Target glands of FSH and LH

17. _____ Cells that surround the oocytes

18. _____ Mature ovarian follicle

19. _____ Ovarian structure that primarily secretes progesterone

20. _____ Called *oviducts* or *uterine tubes*

21. _____ Organ that has borne the entire human race

MATCHING

Structure of the Breasts: Lactation

Directions: Match the following terms to the most appropriate definition by writing the correct letter in the space provided. Some terms may be used more than once. See text, pp. 510, 511.

A. anterior pituitary gland   G. neurohypophysis
B. suspensory ligaments   H. oxytocin
C. areola   I. colostrum
D. lactiferous ducts   J. milk let-down
E. lactation   K. pectoralis major
F. lactogenic hormone   L. alveolar glands

1. _____ The circular area of pigmented skin that surrounds the nipple

2. _____ This muscle is located deep to the breasts.

3. _____ Another name for milk production

4. _____ Reflexive response to the sucking of the infant at breast

5. _____ Name of the milk-secreting glands

6. _____ A yellowish, watery fluid, rich in protein and antibodies; secreted by the mammary glands for 2–3 days after delivery

7. _____ Another name for prolactin

8. _____ Gland that secretes prolactin

9. _____ The gland that secretes oxytocin

10. _____ Sucking by the infant at breast is the stimulus for the secretion of this hormone

11. _____ Milk is secreted into these ducts within the breast.

12. _____ Hormone that is dependent on a hypothalamic releasing hormone

13. _____ Hormone that also causes myometrial contraction

14. _____ Connective tissue that helps support the breasts

15. _____ The hormone that is most associated with the milk let-down reflex

MATCHING

Hormones of the Female Reproductive Cycle

Directions: Match the following terms to the most appropriate definition by writing the correct letter in the space provided. Some terms may be used more than once. See text, pp. 506-510.

A. gonadotropins
B. estrogen
C. releasing hormones
D. LH
E. follicle-stimulating hormone (FSH)
F. progesterone
G. human chorionic gonadotropin (hCG)

1. _____ Promotes the maturation of the egg and the development of the female secondary sex characteristics

2. _____ The "pill" contains estrogen and this hormone.

3. _____ Hormone that is primarily secreted by the ovaries during the follicular phase of the ovarian cycle

4. _____ Hormone that is primarily secreted by the ovaries during the luteal phase of the ovarian cycle

5. _____ A surge of this hormone immediately precedes ovulation.

6. _____ Secreted primarily by the corpus luteum

7. _____ A female looks female primarily because of this hormone.

8. _____ Adenohypophyseal gonadotropin that acts on the ovary to stimulate the development of a follicle

9. _____ Secreted by the hypothalamus

10. _____ Classification of FSH and LH

11. _____ Hormone that stimulates the corpus luteum during early pregnancy and prevents its deterioration

12. _____ When blood levels of estrogen and progesterone decrease at the end of the secretory phase of the uterine cycle, the anterior pituitary secretes this gonadotropin in an attempt to stimulate the development of another egg.

13. _____ Ovarian hormone that dominates the proliferative phase of the uterine cycle

14. _____ Ovarian hormone that dominates the secretory phase of the uterine cycle

MATCHING

Cycles of the Female Reproductive System

Directions: Match the following terms to the most appropriate definition by writing the correct letter in the space provided. Some terms may be used more than once. See text, pp. 506-510.

A. ovarian cycle
B. uterine cycle
C. follicular phase
D. luteal phase
E. ovulation
F. menstrual phase

1. _____ Composed of the follicular phase and luteal phase

2. _____ Also called the *period*

3. _____ Composed of the menstrual, proliferative, and secretory phases

4. _____ The ovaries primarily secrete estrogen during this phase of the ovarian cycle.

5. _____ The ovaries primarily secrete progesterone during this phase of the ovarian cycle.

6. _____ Ejection of a mature egg at midcycle

7. _____ Bleeding characterizes the menstrual phase of this cycle.

8. _____ The secretory phase of this cycle is primarily caused by the secretion of progesterone.

9. _____ The proliferative phase of this cycle is primarily caused by the secretion of estrogen.

READ THE DIAGRAM

Female Reproductive Tract

Directions: Referring to the illustration, indicate the parts of the female reproductive tract by writing the numbers on the lines provided below. Some terms may be used more than once. See text, p. 512.

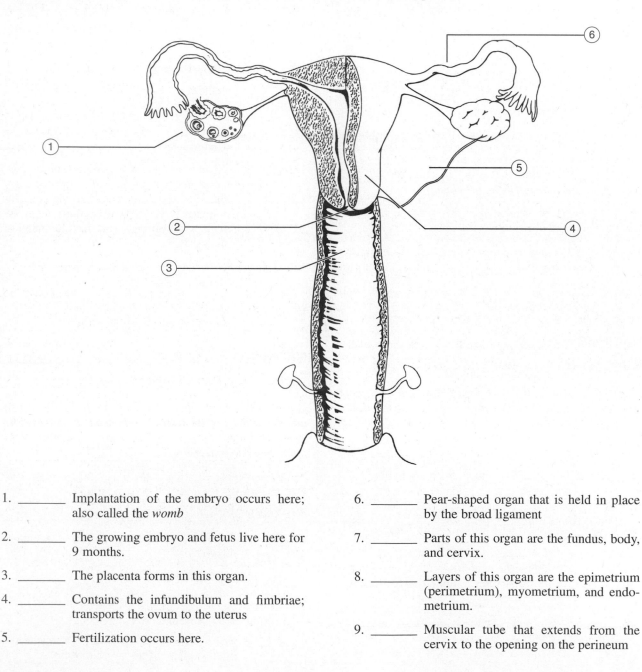

1. _____ Implantation of the embryo occurs here; also called the *womb*

2. _____ The growing embryo and fetus live here for 9 months.

3. _____ The placenta forms in this organ.

4. _____ Contains the infundibulum and fimbriae; transports the ovum to the uterus

5. _____ Fertilization occurs here.

6. _____ Pear-shaped organ that is held in place by the broad ligament

7. _____ Parts of this organ are the fundus, body, and cervix.

8. _____ Layers of this organ are the epimetrium (perimetrium), myometrium, and endometrium.

9. _____ Muscular tube that extends from the cervix to the opening on the perineum

10. _____ Female gonad

11. _____ Broad ligament

12. _____ Cervix of the uterus

13. _____ Target gland of FSH and LH

14. _____ Called the *oviduct* or *uterine tube*

15. _____ Home of the corpus luteum

16. _____ Gland that secretes estrogen and progesterone

17. _____ Structure that ovulates

18. _____ Target of hCG

COLORING AND Xs AND Os

Directions: Color or mark the appropriate areas on the illustration on the previous page as indicated below.

1. Color the corpus luteum *yellow*.

2. Draw (hatched line) the path that the sperm follows to reach the egg for fertilization *blue*.

3. Place an *X* where fertilization normally occurs.

4. Place an *O* where implantation normally occurs.

5. There is a gonococcus in the vagina. Draw *arrows* showing the path that the pathogen can follow on its way to the peritoneal cavity.

SIMILARS AND DISSIMILARS

Directions: Circle the word in each group that is least similar to the others. Indicate the similarity of the three words on the line below each question.

1. fundus   cervix   corpus luteum   body

_____

2. ovary   corpus luteum   graafian follicle   mammary gland

_____

3. gonadotropin   oxytocin   FSH   LH

_____

4. estrogen   progesterone   ovarian hormones   semen

_____

5. cervix   fimbriae   fallopian tubes   oviducts

_____

6. corpus luteum   corpus albicans   graafian follicle   mammary glands

_____

7. acrosome   oocyte   head   flagellum

_____

8. vas deferens   ureter   ejaculatory duct   urethra

_____

9. corpus luteum   seminiferous tubules   prostate gland   vas deferens

_____

10. milk production   lactogenic hormone   prolactin   ovulation

_____

11. endometrial   myometrial   uterine   ovarian

_____

12. FSH   LH   ICSH   testosterone

_____

13. erection   emission   ovulation   ejaculation

_____

14. androgen   testosterone   virilization   ovulation

_____

Part II: Putting It All Together

MULTIPLE CHOICE

Directions: Choose the correct answer.

1. If fertilization does not occur, the corpus luteum
   a. secretes progesterone.
   b. secretes hCG.
   c. forms the corpus albicans.
   d. forms into a placenta.

2. Ovulation
   a. normally occurs on day 14 of a 28-day cycle.
   b. is caused by the secretion of hCG by the tropho-blastic cells.
   c. causes the release of hormones that form the graafian follicle.
   d. normally occurs on day 28 of a 28-day cycle.

3. When is fertilization most likely to occur?
   a. immediately after menstruation
   b. immediately before menstruation
   c. around ovulation (midcycle)
   d. day 28 of a 28-day cycle

4. With which event is the LH surge most associated?
   a. implantation
   b. ovulation
   c. ejaculation
   d. lactation

5. The luteal phase of the ovarian cycle
   a. is dominated by progesterone.
   b. requires hCG.
   c. occurs only in the pregnant state.
   d. immediately precedes ovulation.

6. The endometrium, myometrium, and perimetrium are
   a. embryonic germ layers.
   b. layers of the uterus.
   c. segments of the fallopian tubes.
   d. uterine glands that secrete the hormones of pregnancy.

7. The seminal vesicles, bulbourethral glands, and prostate gland
   a. secrete testosterone.
   b. are called *external genitalia*.
   c. contribute to the formation of semen.
   d. are sites of sperm production.

8. What is the cause of an erection?
   a. secretion of testosterone
   b. contraction of the ejaculatory ducts
   c. filling of erectile tissue with blood
   d. emission

9. Which of the following results in sterility?
   a. phimosis
   b. circumcision
   c. vasectomy
   d. emission

10. Which of the following word-pairs best describes why it is necessary that the testes descend into the scrotum?
    a. temperature, fertility
    b. infection, sterility
    c. cyanosis, ischemia
    d. jaundice, implantation

11. The shaft, or body, of this structure contains three columns of erectile tissue.
    a. epididymis
    b. vas deferens
    c. urethra
    d. penis

12. Which of the following is true of the process of meiosis?
    a. occurs in every cell in the body except the oocyte and spermatogonium
    b. reduces the chromosome number from 46 to 23
    c. increases the chromosome number from 23 to 46
    d. occurs in the ovum but not in the sperm

13. Which of the following is true of menopause?
    a. usually occurs in the late 20s and early 30s
    b. is always accompanied by severe and debilitating symptoms
    c. is accompanied by a gradual decrease in menstrual periods
    d. is accompanied by an increase in estrogen and progesterone secretion

14. What is the term that refers to the implantation and growth of the fetus into the walls of the fallopian tubes?
    a. menses
    b. menarche
    c. ectopic pregnancy
    d. ovulation

15. What is the effect of suckling at breast and oxytocin release?
    a. parturition
    b. implantation
    c. milk let-down reflex
    d. ovulation

16. Which structure secretes colostrum?
    a. testes
    b. mammary glands
    c. prostate gland
    d. graafian follicle

17. Which group is incorrect?
    a. parts of a sperm: head, body, corpus luteum
    b. parts of the uterus: fundus, body, cervix
    c. layers of the uterus: perimetrium, myometrium, endometrium
    d. phases of the uterine cycle: menstrual phase, proliferative phase, secretory phase

18. Which group is incorrect?
    a. layers of the uterus: endometrium, myometrium, graafian follicle
    b. genital ducts: epididymis, ejaculatory ducts, vas deferens, urethra
    c. glands: seminal vesicles, bulbourethral, prostate
    d. phases of the ovarian cycle: follicular phase, luteal phase

220

19. Which group is incorrect?
    a. glands: seminal vesicles, bulbourethral, prostate
    b. phases of the uterine cycle: menstrual phase, luteal phase, secretory phase
    c. gonads: ovaries, testes
    d. external genitals: penis, scrotum, labia, mons pubis

CASE STUDY

P.R. has just given birth to a 7-lb baby girl and intends to breast-feed. Even though her milk has not yet come in, P.R. is encouraged to nurse her baby frequently.

1. How will the baby be nourished during the first few days?
    a. She will receive no nourishment while nursing.
    b. She has no appetite and therefore will not nurse.
    c. She will receive nourishment from the colostrum.
    d. She will receive nourishment from bottle feeding until her mother's milk comes in.

2. What is accomplished by frequent nursing or suckling?
    a. suppresses the secretion of prolactin
    b. stimulates lactation and the flow of milk
    c. suppresses oxytocin
    d. causes uterine bleeding

PUZZLE

Hint: "Don't Let It Die"

Directions: Perform the following functions on the Sequence of Words that follows. When all the functions have been performed, you are left with a word or words related to the hint. Record your answer in the space provided.

Functions: Remove the following:

1. FSH and LH

2. Its midcycle surge causes ovulation.

3. Two ovarian hormones

4. Hormone that stimulates myometrial contraction and is involved in the milk let-down reflex

5. Phases of the ovarian cycle (two)

6. Phases of the uterine cycle (three)

7. Stimulates the breast tissue to make milk

8. Site of fertilization

9. Layers of the uterus

Sequence of Words

G O N A D O T R O P I N S M Y O M E T R I U M S E
C R E T O R Y E S T R O G E N L U T E A L P R O L I
F E R A T I V E P R O L A C T I N P R O G E S T E R
O N E L H O X Y T O C I N F O L L I C U L A R M E
N S T R U A L E N D O M E T R I U M H U M A N C
H O R I O N I C G O N A D O T R O P I N F A L L O
P I A N T U B E M A I N T A I N S T H E C O R P U S
L U T E U M E P I M E T R I U M

Answer: _____

BODY TOON

Hint: Another Name for Ms. Fallopian's Pet

Answer: oviduck

Answer Key: Textbook page references are provided as a guide for answering these questions. A complete answer key was provided for your instructor.

OBJECTIVES

1. Describe the process of fertilization: when, where, and how it occurs.
2. Do the following regarding prenatal development:
   - Describe the process of development: cleavage, growth, morphogenesis, and differentiation.
   - Explain the three periods of prenatal development: early embryonic, embryonic, and fetal.
   - State two functions of the placenta.
3. Explain hormonal changes during pregnancy.
4. Describe the hormonal changes of labor.
5. Describe immediate postnatal changes and lifelong developmental stages.
6. Discuss heredity and how genetic structures are related, including:
   - Describe the relationships among deoxyribonucleic acid (DNA), chromosomes, and genes.
   - Define *karyotype*.
7. Explain how the gender of the child is determined.
8. State the difference between congenital and hereditary diseases.

Part I: Mastering the Basics

MATCHING

Fertilization and Embryonic Development

Directions: Match the following terms to the most appropriate definition by writing the correct letter in the space provided. Some terms may be used more than once. See text, pp. 519-522.

A. fertilization   E. cleavage
B. zygote   F. morphogenesis
C. pregnancy
D. differentiation

1. _____ A zygote-making event

2. _____ Developmental process that refers to the shaping of the cell cluster

3. _____ Refers to the union of the nuclei of the egg and sperm

4. _____ Fertilized egg

5. _____ A cell undergoes this developmental process to become a specialized cell, such as a nerve cell, muscle cell, or blood cell.

6. _____ Process that restores the haploid number (23 chromosomes) to the diploid number (46 chromosomes)

7. _____ Time of prenatal development that is also called *gestation*; divided into trimesters

8. _____ Also called *conception*

9. _____ Mitosis: 4, 8, 16 cells

MATCHING

Early Embryonic Period to Embryonic Period

Directions: Match the following terms to the most appropriate definition by writing the correct letter in the space provided. Some terms can be used more than once. See text, pp. 519-527.

A. implantation   I. amniotic fluid
B. inner cell mass   J. primary germ layers
C. morula   K. amniotic sac
D. embryo   L. blastocyst
E. blastomeres   M. extraembryonic
F. zygote   membranes
G. chorion   N. placenta
H. umbilical cord

1. _____ Process whereby the late blastocyst burrows into the endometrial lining of the uterus

2. _____ Through cleavage, the zygote is transformed into this raspberry-shaped cluster of 16 cells.

3. _____ Cluster of cells of the blastocyst that forms the embryonic disc

4. _____ Called the *bag of waters*; this bag breaks before delivery and generally signals the onset of labor.

5. _____ Amnion, chorion, yolk sac, and allantois

6. _____ Found within the amniotic sac; it forms a cushion around the embryo that helps protect it from bumps and changes in temperature

7. _____ Name of baby-to-be from week 3 to week 8

8. _____ The two-cell, four-cell, and eight-cell clusters during early embryonic development

9. _____ About the fifth day, the morula becomes this.

10. _____ After delivery, the stump of this structure becomes the navel, or belly button.

11. _____ Ectoderm, mesoderm, endoderm; all the tissues and organs of the body develop from these structures

12. _____ The outer extraembryonic membrane that forms finger-like projections called *villi*; helps form the placenta

13. _____ This disc-shaped structure is the site where the fetal and maternal blood circulations meet; baby eats, excretes, and breathes at this site.

14. _____ The hook-up or lifeline between the baby and the placenta; contains blood vessels

15. _____ Sperm meets egg and is called a(n) _____.

READ THE DIAGRAM

Directions: Refer to Fig. 27.1 in the textbook, and place the following events in the correct sequence (Column 1). Indicate in Column 2 where the event takes place.

late blastocyst	implantation
fertilization	ovulation
morula formation	early blastocyst
blastomere formation	

Column 1 *(Ordering)*	*Column 2* *(Location)*
_____	_____
_____	_____
blastomere formation	fallopian tube
_____	_____
_____	_____
_____	_____
_____	_____

MATCHING

Fetal Period and Birth

Directions: Match the following terms to the most appropriate definition by writing the correct letter in the space provided. See text, pp. 526-529.

A. fetus	F. quickening
B. vernix caseosa	G. lanugo
C. Braxton-Hicks contractions	
D. labor	
E. parturition	

1. _____ The developing offspring from week 9 to birth

2. _____ The mother's experience of first feeling the fetus move during the fifth month

3. _____ The birth process

4. _____ Stages are dilation, expulsion, and placental

5. _____ Weak, ineffectual uterine contractions that normally occur during late pregnancy; often associated with false labor

6. _____ A fine downy hair that covers the fetus; appears during the fifth month

7. _____ Process whereby forceful contractions expel the fetus from the uterus

8. _____ White, cheeselike substance that protects the fetus' skin from the amniotic fluid; secreted by the sebaceous glands of the fetus

MATCHING

Hormones of Pregnancy

Directions: Match the following terms to the most appropriate definitions by writing the correct letter in the space provided. Some terms may be used more than once. See text, pp. 520-524, 525-529.

A. human chorionic gonadotropin (hCG)	D. progesterone
	E. aldosterone
B. oxytocin	
C. parathyroid hormone (PTH)	

1. _____ Secreted by the trophoblastic cells of the blastocyst during implantation

2. _____ The placenta takes over the role of the corpus luteum and secretes estrogen and this hormone.

3. _____ Throughout pregnancy, this hormone inhibits uterine contractions.

4. _____ Thought to be responsible for morning sickness during the first trimester

5. _____ Posterior pituitary hormone that stimulates uterine contractions; plays an important role in labor

6. _____ Hormone that is the basis of the pregnancy test

7. _____ Hormone that is released by the adrenal cortex; stimulates the kidneys to reabsorb Na^+ and water, thereby expanding the maternal blood volume

8. _____ Hormone that keeps the maternal plasma levels of calcium high; provides baby with adequate calcium for bone growth

9. _____ The corpus luteum secretes some estrogen and larger amounts of this hormone.

10. _____ This hormone prevents deterioration of the corpus luteum.

MATCHING

DNA, Genes, Chromosomes

Directions: Match the following terms to the most appropriate definition by writing the correct letter in the space provided. Some terms may be used more than once. See text, pp. 529-533.

A. DNA
B. genes
C. chromosomes
D. autosomes
E. codominant gene
F. recessive gene
G. Down syndrome
H. mutation
I. nondisjunction
J. karyotype
K. carrier
L. dominant gene
M. sex-linked trait
N. sex chromosomes

1. _____ Nucleotides that contain the base sequencing (code) for genetic information

2. _____ DNA is tightly wound into these tightly coiled, threadlike structures.

3. _____ Segments of a DNA strand that carry the code for a specific trait, such as skin color or blood type

4. _____ Failure of the chromosomes (or chromatids) to separate during meiosis, thereby causing the formation of eggs or sperm that have too many chromosomes

5. _____ There are 23 pairs or 46 of these in almost every human cell.

6. _____ X and Y chromosomes

7. _____ Genes that express a trait equally (e.g., AB blood type)

8. _____ Any trait that is carried on an X or a Y chromosome

9. _____ Child who has three copies of chromosome 21

10. _____ Type of gene that is not expressed if it is paired with a dominant gene

11. _____ Process of meiosis reduces the numbers of these by half (from 46 to 23).

12. _____ 22 pairs (numbered 1 to 22) of the chromosomes

13. _____ Arrangement of chromosomes by size and shape; genetic art

14. _____ Type of gene that expresses itself (e.g., brown eyes are expressed over blue eyes)

15. _____ Change in the genetic code that may express itself by a change in a particular trait

16. _____ Person who shows no evidence of a trait (like blue eyes) but has a recessive gene for that trait

SIMILARS AND DISSIMILARS

Directions: Circle the word in each group that is least similar to the others. Indicate the similarity of the three words on the line below each question.

1. zygote formation   fertilization

   implantation   conception

_____

2. zygote   placenta   morula   blastocyst

_____

3. pregnancy   ovulation   gestation   three trimesters

_____

4. trophoblastic cells   corpus luteum   hCG   prolactin

_____

5. gonad   twins   monozygotic   dizygotic

_____

6. amnion   graafian follicle   chorion   allantois

_____

7. placenta   lacunae   oviduct   chorionic villi

_____

8. umbilical vein   umbilical cord   embryonic disc   umbilical arteries

_____

9. parturition   myometrial   conception   birth
conception

_____

10. oxytocin   prolactin   testosterone   lactogenic
hormone

_____

11. dominant   congenital   recessive   codominant

_____

12. sex chromosomes   XX   XY   autosomal
chromosomes

_____

Part II: Putting It All Together

MULTIPLE CHOICE

Directions: Choose the correct answer.

1. Fertilization
   a. usually occurs on day 1 of a 28-day cycle.
   b. occurs within the fallopian tube.
   c. results in formation of the graafian follicle.
   d. is a direct response to follicle-stimulating hormone (FSH) and luteinizing hormone (LH) stimulation.

2. When you were a zygote, how many cells did you have?
   a. 1
   b. 4
   c. 8
   d. 16

3. How many cells did you have when you looked most like a raspberry (morula)?
   a. 1
   b. 4
   c. 8
   d. 16

4. What is another term for a spontaneous abortion?
   a. induced abortion
   b. eclampsia
   c. placenta previa
   d. miscarriage

5. Which of the following occurs last?
   a. morula formation
   b. fertilization
   c. zygote formation
   d. implantation

6. Which of the following terms implies a genetic problem?
   a. teratogen
   b. hereditary
   c. congenital
   d. fetal alcohol syndrome

7. Which of the following conditions is most related to fetal alcohol syndrome?
   a. infection
   b. rubella
   c. teratogen
   d. heredity

8. Which of the following statements is related to the early embryonic period, the embryonic period, and the fetal period?
   a. stages of labor
   b. prenatal development
   c. first trimester of pregnancy
   d. periods of organogenesis

9. Which of the following is true of monozygotic twins?
   a. Each twin develops from a different zygote.
   b. They are fraternal twins.
   c. They are identical twins.
   d. They have congenital defects.

10. Which period involves the formation of extraembryonic membranes, the placenta, and the period of organogenesis?
    a. period of implantation
    b. blastomere stage
    c. third trimester
    d. embryonic period

11. Which of the following is a correct combination?
    a. fertilization—ovary
    b. implantation—uterus
    c. morula formation—cervix
    d. corpus luteum—oviducts

12. Which of the following describes the organism immediately after birth?
    a. fetus
    b. embryo
    c. blastocyst
    d. neonate

13. Which of the following is true of the fetal heart?
    a. the last organ to form
    b. does not become functional until the sixth month
    c. present by the second month but does not begin pumping blood until birth when the fetal heart structures disappear
    d. pumps blood as early as the second month

14. Teratogens
    a. are fetal heart structures.
    b. cause hereditary birth defects.
    c. are always infectious agents.
    d. cause congenital defects.

15. At birth, the extraembryonic membranes and the placenta are expelled as which structure?
    a. corpus albicans
    b. morula
    c. myometrium
    d. afterbirth

16. An oxytocic agent is one that stimulates which structure?
    a. fallopian tubes
    b. myometrium
    c. ovaries
    d. vagina

17. What is true about a child with an X and a Y chromosome?
    a. The child has Down syndrome.
    b. The child is male.
    c. The child is female.
    d. The child has ovaries.

18. Which group is incorrect?
    a. extraembryonic membranes: amnion, chorion, yolk sac, allantois
    b. primary germ layers: ectoderm, mesoderm, endoderm
    c. embryonic structures: zygote, blastomere, morula, blastocyst, fetus
    d. genes: dominant, recessive, codominant

19. Which group is incorrect?
    a. embryonic structures: zygote, blastomere, morula, blastocyst
    b. hormones: estrogen, progesterone, hCG, prolactin, oxytocin
    c. genetic disorders: Down syndrome, Edward's syndrome, eclampsia
    d. primary germ layers: ectoderm, mesoderm, endoderm

20. Which group is incorrect?
    a. genes: dominant, recessive, codominant
    b. genetic disorders: Down syndrome, Edward's syndrome, Patau syndrome
    c. fetal characteristics: lanugo, quickening, milk let-down reflex
    d. ovarian structures: corpus luteum, corpus albicans, graafian follicle

CASE STUDY

A 22-year-old gave birth to a healthy baby girl. She noticed that her baby looked like she was covered by a white cream cheeselike substance and had fine hair over a large area of her body.

1. Which of the following best describes the appearance of her baby?
    a. quickening, implantation
    b. vernix caseosa, lanugo
    c. vernix caseosa, quickening
    d. Braxton-Hicks, milk let-down

2. Her newborn infant is best described as a(n)
    a. fetus.
    b. embryo.
    c. newborn with XY chromosomes.
    d. neonate.

Hint: "A Zygoting Event"

Directions: Perform the following functions on the Sequence of Words that follows. When all the functions have been performed, you are left with a word or words related to the hint. Record your answer in the space provided.

Functions: Remove the following:

1. Structure in which fertilization normally occurs

2. Organ in which implantation normally occurs

3. Distal necklike region of the uterus

4. Muscle layer of the uterus

5. Cells that secrete hCG

6. Top dome-shaped part of the uterus

7. This gland also serves as the site where the fetus breathes, eats, and excretes.

8. Fluid in which the fetus is immersed

9. Extraembryonic membranes (four)

10. A virus or drug that is capable of causing a monster-like congenital defect

11. Three primary germ layers

12. Baby is covered in white cream cheese

13. Organ that produces the mature ovum

14. The blastomere gives rise to this raspberry.

15. Types of twins

Sequence of Words

F A L L O P I A N T U B E O V A R Y M E S O D E R
M A M N I O N P L A C E N T A A M N I O T I C F U
N D U S C H O R I O N E N D O D E R M M Y O M E
T R I U M T R O P H O B L A S T I C C E L L S Y O L
K S A C D I Z Y G O T I C V E R N I X C A S E O S A C
E R V I X U T E R U S T E R A T O G E N E C T O D E
R M M O N O Z Y G O T I C A L L A N T O I S F E R T
I L I Z A T I O N M O R U L A

Answer: _____